The Brain, the Mind and

Psychoanalysis enjoyed an enormous popularity at one time, but has recently fallen out of favor as new psychiatric medications have dominated the treatment of mental illness and a new interest in the brain and neuroscience begins to dominate the theory as to the cause and cure of mental illness.

How do we distinguish between the brain, the mind and the self? In his new book, Arnold Goldberg approaches this question from a psychoanalytic perspective, and examines how recent research findings can shed light on it. He repositions psychoanalysis as an interpretive science that is a different activity to most other sciences that are considered empirical.

Giving clear coverage of the various psychoanalytic models of the mind and the self, Goldberg examines how these theories fare against neuroscientific evidence, and what implications these have for psychoanalytic clinical practice. *The Brain, the Mind and the Self: A psychoanalytic road map* sets up evidence-based, robust psychoanalytic theory and practice that will give psychoanalysts, social workers and practicing psychologists a valuable insight into the future of psychoanalysis.

Arnold Goldberg, M.D. was born and raised in Chicago and trained at the University of Illinois, Michael Reese Hospital and the Institute for Psychoanalysis in Chicago. He is recently retired from the Cynthia Oudejans Harris MD chair, and Professor of Psychiatry at Rush Medical Center.

The Brain, the Mind and the Self

A psychoanalytic road map

Arnold Goldberg

LONDON AND NEW YORK

First published 2015
by Routledge
27 Church Road, Hove, East Sussex, BN3 2FA

And by Routledge
711 Third Avenue, New York, NY 10017

*Routledge is an imprint of the Taylor & Francis Group,
an informa business*

© 2015 Arnold Goldberg

The right of Arnold Goldberg to be identified as author of this work has been asserted by him in accordance with sections 77 and 78 of the Copyright, Designs and Patents Act 1988.

All rights reserved. No part of this book may be reprinted or reproduced or utilised in any form or by any electronic, mechanical, or other means, now known or hereafter invented, including photocopying and recording, or in any information storage or retrieval system, without permission in writing from the publishers.

Trademark notice: Product or corporate names may be trademarks or registered trademarks, and are used only for identification and explanation without intent to infringe.

British Library Cataloguing in Publication Data
A catalogue record for this book is available from the British Library

Library of Congress Cataloging-in-Publication Data
Goldberg, Arnold, 1929– , author.
  The brain, the mind and the self : a psychoanalytic road map / Arnold Goldberg.
     p. ; cm.
  I. Title.
  [DNLM: 1. Psychoanalysis. 2. Psychoanalytic Theory. 3. Mental Processes. 4. Neurosciences. WM 460.2]
  RC504
  616.89'17—dc23
  2014042825

ISBN: 978-1-138-78832-9 (hbk)
ISBN: 978-1-138-78833-6 (pbk)
ISBN: 978-1-315-76561-7 (ebk)

Typeset in Times
by Apex CoVantage, LLC

"The essence of something is not at all to be discovered simply like a fact; on the contrary, it must be brought forth, since it is not directly present in the sphere of immediate representing and intending. To bring forth is a kind of making, and so there resides in all grasping and positing of the essence: something creative."

Martin Heidegger

Contents

Introduction   1

PART I
Distinguishing the brain, the mind and the self   5

1 The brain, the mind and the self: three conundrums in psychiatry and psychoanalysis   7

2 On understanding "understanding": how we understand the meaning of the word "understanding"   17

3 On the scientific status of psychoanalysis and psychotherapy   29

4 A psychoanalysis independent of psychiatry   39

5 The concepts of normality and psychopathology in psychoanalysis   53

PART II
The newer models of the mind and the self   67

6 Being kept in mind   69

7 The many meanings of empathy   77

8 Self-empathy   101

**PART III
Clinical examples of the special role
of psychoanalysis** 109

9 The danger in diversity 113
10 Reflections on enthusiasm 123
11 On outrage and the need to be mad 135
12 Carving out a place for psychoanalysis 143
13 The future: epilogue 151

Bibliography 153
Index 161

Introduction
The brain, the mind and the self: a psychoanalytic road map

Inasmuch as I have lived through the heyday of psychoanalysis and am now experiencing what seems to be its ultimate decline, an autobiographical account seems both permissible and warranted. Those of us who decided to pursue psychiatric training after medical school were united in the view that becoming a psychoanalyst was our ultimate goal. Our training during psychiatric residency was primarily based on treating patients with psychoanalytic concepts of treatment. The few somatic treatments, such as electroshock therapy or insulin comas that were available were reserved for the "inpatients" who were usually psychotic and beyond "talk therapy." There were no drugs, save the barbiturates, until Thorazine came upon the scene – soon to be followed by legions of medications. Psychiatrists who eschewed psychoanalysis seemed primarily to care for these very sick inpatients and resorted mainly to these few somatic therapies. Some psychoanalysts, however, did choose to apply psychoanalytic ideas to psychoses, and long hospitalizations of over a year were commonplace. Although the change was slow, it was dramatic, albeit mightily ignored by many psychiatrists; drug therapy, changed everything. Beginning with the anti-psychotics and followed by the anti-anxiety drugs and antidepressants, a new breed of psychiatrist evolved, and with this rather noteworthy historical movement, psychoanalysis as practiced in the training of psychiatrists began its decline. No doubt many psychoanalysts maintained a stalwart faith in what they did and how they did it, but overall this became a vanishing breed.

The medical student who chose to become a psychiatrist was once the primary candidate or student to enter a psychoanalytic institute in order to become a psychoanalyst, but now fewer physicians make this choice, and alongside the new institutes that trained non-medical students, the older institutes began accepting social workers and psychologists for training. Concurrent with the decline in the usual form of psychoanalytic treatment, there arose a somewhat watered-down form of therapy initially termed psychodynamic therapy. This was distinguished from what was called "classical" psychoanalysis by being less frequent and less lengthy, as well as less costly. It may have started from the efforts to treat patients with psychoanalytic ideas by those who were not yet full-fledged psychoanalysts, but ultimately it became an alternative to psychoanalysis proper and soon dominated all "talk therapy."

The status of psychoanalysis changed markedly as alternative treatments came upon the scene. Whereas once the chairman of the department of psychiatry at a medical school was almost required to be a psychoanalyst, the office paradigm transformed to his or her needing to be an expert in psychotropic medications, with analytic training either being unnecessary or even a hindrance. Fewer and fewer people, either medical or non-medical, applied to psychoanalytic institutes for training, and a host of other forms of treatment ranging from cognitive behavioral therapy to mindfulness came to be popular both because of time and cost, as well as effectiveness for a variety of patients. Psychoanalysis began to appear almost "quaint."

The decline of psychoanalysis and the upsurge in drug therapy for mental illness shifted the basic theoretical study from that of the workings of the mind to the functions of the brain, and slowly the theory of the mind became that of the brain. A recent book by an eminent computer expert is entitled *How to Create a Mind: The Secret of Human Thought Revealed* (Kurzweil, 2012), even though it is entirely about the brain. It is quite representative of present-day thinking, which assumes that the organ inside the skull is where the mind is and so assumes that the mind is just another word for the brain.

Psychoanalysis reacted to this new focus on the brain (now called neuroscience) with a combination of responses. Some analysts ignored it, some embraced it, and some worked to incorporate it into a collaborative effort. However, over the years psychoanalysis has itself changed, primarily by dividing into a number of groups usually heralding a single person who concentrates on one or another variant of psychoanalysis. There are thus a series of forms of psychoanalysis ranging from Melanie Klein and followers of Kleinian thought to those championing Jacques Lacan, Heinz Kohut, Wilfred Bion and on and on to many groups hardly known outside of their particular geographic location. Such splintering or divisiveness is often applauded as a sign of "pluralism," and this pluralism can lead to specialized vocabularies and a lack of agreement as to what psychoanalysis is. This erosion of a core definition of psychoanalysis seems to parallel its claim as to its being a proper study of the mind.

The debate of mind versus brain was unable to restrict itself to these two concepts when a new version of an old term came to claim prominence in the form of the self. Although ignored by some neuroscientists as irrelevant to a localization in the brain, a few did attempt to find a place for it in that organ with one almost laughable conclusion that likened the self to "the center of gravity" (Dennett, 2013a), thus allowing it a status without a location. Regardless of that, the position of the self began to be a heated controversy within psychoanalysis.

Over time, the argument against equating the brain with the mind was taken up primarily by philosophers, along with some psychologists. Psychoanalysis was either unwilling or unable to participate in the debate, so preoccupied was it with its own diminishing significance. Furthermore, squabbles over "pluralism" began to appear more and more as problems of the "silo effect," which, interestingly, also started to appear in psychiatry. The "silo effect" is a term applied to insulated areas of expertise and activity that guard themselves against the intrusion of other areas of expertise and activity. Thus, those who are proficient in medication trumpet their own work and ignore psychodynamics, while those proficient in psychodynamics ignore cognitive behavioral therapy, while those proficient in cognitive behavioral therapy ignore medication, and on and on. Although not true of any single analyst or therapist, the "silo effect" does seem to be an accurate, although perhaps unfortunate, result of becoming proficient – namely, emphasizing exclusively whatever it is that you do best.

With the field of psychiatry focusing initially on the brain and soon thereafter upon the genetic basis of mental disorders, it became clear that it had less and less in common with psychoanalysis and was not at all interested in these various schools of psychoanalysis. Indeed, although Sigmund Freud was both neurologist and psychiatrist, there no longer seemed to be any good reason for this linkage, save the historical one. Although some patients came to find treatment with a person trained and proficient in both psychodynamic or psychoanalytic therapy as well as psychotropic medication, this combination became rarer and rarer, and there was no clear and definitive determination as to what was best for what malady. It was not like determining which antibiotic worked best for what infection but more like prescribing a pain medication without diagnosing the origin of the pain. Some therapists treated the brain. Some treated the mind. Some treated the self. Psychiatry was excited about the possibilities of finally finding the cause of mental illness either in the genes or in a particular part of the brain. Psychoanalysis struggled with efforts to contain pluralism while reaching out to the humanities and non-medical applicants. As these two fields grew further apart, a host of therapeutic efforts formed to care for those who chose neither psychoanalysis *nor* medication. When the mayor of a city was found to be sexually mistreating his employees, he agreed to enter a two-week course of treatment to cure his supposed mental illness. The treatment was cut short for unknown reasons, but it nicely represented the present state of disarray in today's therapeutic approach to problems of the mind. It certainly begins with our disagreement on just what the mind may be. This book is a start in that direction. It purports to clarify the many uses and abuses of certain psychological terms. At its heart is the claim that the brain is essential for all human action and thought. However, what the self does is not what the brain does, and what is on or in your mind is not on or in the brain. It is not merely a problem with language as much as it is a problem of meaning. Only, or better said, psychoanalysis is the main field that clarifies the distinctions between brain, mind and self. That is where we now turn.

Part I

Distinguishing the brain, the mind and the self

Chapter 1

The brain, the mind and the self

Three conundrums in psychiatry and psychoanalysis

The visitor to the field of mental health is first directed to the discipline of psychology and then to the separate area of psychopathology and is soon confronted with what seems to be a battleground of language. If he or she enters the realm of psychiatry, the issues are clear: it is all about the brain and neuroscience. If a turn to psychoanalysis is taken, then things get a bit foggy: psychoanalysis is all about the mind, with some concern with the self, both of which are reasonable concepts. However, a multitude of choices are offered as to theories and techniques dealing with them, peppered with a number of names of supposedly famous persons. The visitor is not allowed a chance to deliberate, since there is an accompanying pressure to commit before deliberation with the warning that one cannot freely comprehend the issues without a full commitment. Also, the visitor cannot for long be a visitor without an allegiance that ultimately transforms one into a believer.

It should come as no surprise. The believers in the brain feel psychodynamics and psychoanalysis are quaint relics that are mainly passé, save for a few disorders that are yet to be managed. The believers in psychoanalysis are a bit more cordial to neuroscience, but feel it cannot ever do the work of analysis. That cordiality does not, however, dominate the field – some believers in psychoanalysis are totally dismissive of the brain, along with all other schools of theory and technique. It is truly a battleground. As such, it is directed and devoted to a single winner with but a few hopeful efforts at reconciliation. Rather, the road to belief is one of the removal of discords. It bears repeating: the brain generates the mind, and all problems are ultimately to be found in and treated in that organ in the skull. The mind is everything that has meaning, and a concentration on the brain is but a diversion. The self is the agent or the person that is involved in and with the world and cannot or should not be reduced to an organ or divorced from the world. That is one solution.

The dictionary defines conundrum as a puzzle or a hard question, and one needs to recognize that such puzzles are often dismissed rather than solved. One effort that is directed toward a solution is that of treating the brain, the mind and the self as three different but related areas of inquiry with no need to reduce them all to a single study, yet with the added recognition that they study what may be or seems to be the same or similar problems. Mistakes occur when we choose to misapply or misdirect our efforts, such as trying to locate the self in the brain or examining the meaning of the ventral striatum. Each area of study is restricted by its own special language and its own rules. Not surprisingly, a number of writers and philosophers have both helped and hindered our efforts at clarification of the conundrum.

It is always a bit risky to begin a work that is meant to be read by non-philosophers with the name of Martin Heidegger, one who is often automatically felt to be rather opaque, even to bona fide philosophers. When first suggested some years ago (Goldberg, 2004), the name was linked to that of Heinz Kohut as an odd couple with the premise that understanding was the common ground between these two thinkers (more about that in Chapter 7). Heidegger felt that understanding is the essence of "being in the world" (Goldberg, 2004, p. 205), whereas Kohut felt that we achieve understanding by way of an empathic connection to someone else. Heidegger insisted that the subject does not stand apart and look at the world – indeed, one is always in the world. Kohut felt that the individual person or the self is never alone and apart from others, and that one must consider others to be a necessary part of the self; for this he coined the word "selfobject."

We now introduce a third person into our effort to initiate a definition and clarification of the three concepts that seem to bedevil the three fields of psychiatry, psychology and psychoanalysis. The concepts are the brain, the mind and the self. The third member of our research efforts is Rupert Sheldrake (Sheldrake, 2012), who posits a field theory of mind, which makes the claim that minds extend beyond the brain, and thus our experiences of the world are not merely a replica of our perceptions but rather extend out far beyond and are influenced by our expectations of the world plus intentions and memories (Sheldrake, p. 223). Another triad.

Neuroscience teaches us that a variety of stimuli impinge upon our brain, and the various connections within the brain thereupon deliver a readout or printout, which is considered to be the mind. Piaget tells us that until age 10, children do not consider the mind to be confined to the head, but rather that it extends into the world beyond us. However, Crick offers us a truism that he calls an Astonishing Hypothesis: "You . . . are . . . no more than the behavior of a vast assembly of nerve cells and their associated molecules" (Crick, 1994, p. 3). Over time many of us seem to accept this hypothesis and so devote our time and energy to reducing the mind to its biological substrate with the accompanying hope that ultimately all mental illness will be managed and treated by a proper study of the brain.

The very idea of a mind extending beyond the brain is usually handled by invoking our imagination, and we all can imagine what things might be like in another time or place, or even by entertaining a number of scenarios that might take place. However, we can readily attribute those exercises to the brain. I first began to appreciate this new idea with the very interesting and peculiar phenomenon of "the stare." Almost everyone has, at one time or another, felt that someone whom they could not see was staring at them. People do "feel" that they are being watched when they are really being watched, and this has even been experimentally confirmed, including with animals and surveillance cameras (Sheldrake, 2012, pp. 225–226). This "attention at a distance" seems to confirm that minds are not confined to the inside of the brain. We seem to be receiving impressions beyond our ordinary senses. One may extend this idea to problems of telepathy and mind reading by psychic powers, but those points may well divest us from the major issue of recognizing that the mind extends beyond the brain, and the Crick hypothesis is wrong.

The mind must be differentiated from the brain without in any way denying that the brain is the substrate of the mind; i.e., the brain does indeed generate the mind just as single letters combine to make up words, and molecules make up complex proteins. We must next differentiate the self from these other two entities. Heinz Kohut felt the self to be constituted by the individual along with his or her selfobjects, which he subdivided into mirroring, idealizing and twinship selfobjects. These "other" persons allow us to extend beyond our skins and to become parts of a larger community. Indeed, we are sustained by our selfobjects.

For some, the problem surrounding the concepts of brain, mind and self is handled with utmost efficiency by a simple denial of the existence of anything other than the brain. The mind may be allowed a presence by having it be identical to the brain. The self is just willed out of existence. Although most philosophers pay lip service to Wittgenstein's caution that our language determines our thinking, the deniers seem to be able to talk about the person who is doing that thinking without acknowledging his or her actuality. To insist that concepts such as beauty or even hate exist is often handled by likening such talk to the search for the soul. The self or the person either cannot be found (as we shall see soon in a quote by Hume) or is a vague concept that need never rise to the level of something tangible like an honest-to-goodness area of the brain. Rest easy, for we will never find the self in the brain inasmuch as we must recognize that language is the ability to use words rather than to assign words to things. One finds wisdom in the Bible or joy in a novel without ever being able to exactly point to that discovery. The search for the self will not result in a location.

There is an old adage in science: "the absence of evidence is not the evidence of absence." Unfortunately, the following quotations reveal that the evidence sought is never to be found.

One effort to give up on learning what a self is can be found in a book by Nicholas Fearn, which begins with two quotes:

> For my part, when I enter most intimately into what I call myself, I always stumble on some particular perception or other of heat or cold, light or shade, love or hatred, pain or pleasure. I can never catch myself at any time without perception, and can never observe anything but the perception.
> David Hume (quoted in Fearn, 2005, p. 3)

One cannot but wonder exactly who is the "I" who enters and stumbles and cannot catch or observe. Ah, it must be the brain. But wait, now this:

> You enter the brain, through the eye, march up the optic nerve, round and round the cortex, looking behind every mirror, and then, before you know it, you emerge into daylight on the spike of a motor neuron impulse, scratching your head and wondering where the self is.
> Daniel Dennett (quoted in Fearn, 2005, p. 3)

Well the self is clearly not in the brain since Dennett tried and failed to find it there. One might wonder exactly who Dennett is, because he and Hume seem to do a great deal of traveling around without a self. Perhaps he and Hume have other means of transportation or are sneaking in a self without paying the cost of admission. They are outstanding examples of trying to not mean what they mean. They travel alone in a self-contradictory vehicle and will never find what they feel that they are looking for.

Thus we must move beyond that fragile organ within our skull to issues that involve a world that has meaning to us and that we are a part of. There is, of course, a natural resistance by our fellow scientists who claim that our mind and our self are no more than the brain. But surely they can be generated by our brain and still be more than our brain. We engage the world by our understanding, which, for Heidegger, is realized in interpretation and for Kohut is achieved by empathy. Here is where psychoanalysis becomes the primary vehicle to allow us to explain how we live in the world rather than merely look out at the world.

Although it may appear to be lacking in precision, we can imagine a number of scenarios that can aid in the differentiation and definition of brain, mind and self. Imagine a picnic on a beach in which a little boy wanders away from his family. He is not concerned with keeping an eye on them or they with watching him. He wanders far and is soon lost. He believes that they (the mother and/or father) will come for him. A motion picture called *Home Alone* was a popular version of this particular imaginary story. The mother has lost her child from her mind and suddenly remembers and recovers him with anxiety and even terror. The child may or may not assume that he is yet in his mother's mind. Although each brain is generating these thoughts and fears, they are temporarily irrelevant to the story. The mother both has her child now in her mind and is also able to be empathic with his fears. How and when he disappeared from her mind and then reappeared is a matter of speculation, but she is now able to feel and see "what it must be like" for him as well as what it is like for her.

A similar clinical situation occurred for Dr. S., who was treating a depressed male patient who regularly professed suicidal thoughts and plans. Dr. S. found that he simply could not get this patient out of his mind, and wanted to call the patient if only to be reassured that the man was alive. A supervisor told Dr. S. that the patient needed to feel that he was being thought about, because he felt empty and lost if he contemplated a state wherein nobody seemed to care about him. The supervisor felt that Dr. S. was a necessary structural component for the patient; i.e., a selfobject. The supervisor understood the nature of this therapeutic connection and interpreted it to Dr. S. The patient was found to be severely suicidal after visiting one of his many physicians and receiving a variety of potential possible maladies. Dr. S. told the patient that he wanted to accompany him to his next doctor's visit in order to determine the onset of those suicidal thoughts, but also to assuage the doctor's own anxiety. He had clearly over-identified with the patient as he could not properly regulate his empathic involvement, which, however, became manageable with the aid of his supervisor's interpretation.

To recapitulate: as the little boy walked along the beach away from his family, he felt that he was a part of the world and connected to them. Children under 10 do not believe that the world resides inside their heads but rather that it really is as it appears and that we are all connected to one another. The mother, at least momentarily, lost her child from her mind until somehow she is made (perhaps by someone asking about him) to reconnect with him. The child's idealization of the mother may allow him the security of being found or, if this is somewhat pathological, the child's failure of idealization may stimulate his own anxiety at being forgotten. No matter how one constructs this fictional scenario, it is clear that the minds of these two protagonists extend far and wide over the beach. We shall not at this point debate the possibility of the child and mother knowing that they are forgotten or remembered. Much the same applies to Dr. S. and his patient.

We surely use the brain, mind and self in different ways to talk about different ideas. These are semantically distinct discourses and need to be distinguished as such. They need not nor should not be reduced to each other. The patient of Dr. S. was depressed and so his brain had an unusual chemical and electrical activity. He also involved a number of physicians in his life and became a part of their minds much as they occupied his own mind. We can explain both the story of the little boy and his mother as well as that of Dr. S. and his patient with psychoanalytic discussions involving the self.

Arva Nöe says, "we are out of our heads. We are in the world and of it. We are patterns of active engagement with fluid boundaries and changing components. We are distributed" (2009, p. 183). Therefore, from the earlier claim made by Crick that we are "no more than the behavior of a vast assembly of nerve cells and their associated molecules" to our being "patterns of active engagement with fluid boundaries" to our being constituted by our nuclear self plus our selfobjects, we may conclude that any effort to reduce us to one idea or thing will likely prove to be futile. Rather, we must recognize that we employ different languages to talk about brain, mind and self, and each language is true and sufficient within itself. One outstanding philosopher who is devoted to reducing all mental activity to the brain, demonstrating the value of seeing the brain as a computer, does confront the problem of meaning by saying,

> Meaning isn't going to be a simple property that maps easily onto brains and we're not going to find "deeper" facts anywhere that just settle the question of

what a sentence, or a thought, or a belief really means. The best we can do – and it is quite enough – is to find and anchor (apparently) best interpretations of all the . . . data we have.

(Dennett, 2013b, p. 197)

One can only wonder why this renowned philosopher does not recognize that he has moved into a well-studied field – i.e., the science of interpretation – and in so doing entered the study of meaning, which is not a lesser activity than the study of computer activity but is surely a different one.

A recent study suggested that brain scans could become useful for the study of mental disorder, just as ECGs were employed for cardiac pathology and so would eliminate or diminish the participation of the subject in this diagnostic venture. The person need not contribute to the investigation by stating how he or she felt. One could also limit the investigation to the other arena of mind and self and so insist that there is no need for a brain scan. And, of course, the psychoanalyst may primarily investigate the unconscious determinates of the mind in a similar single-minded effort. Factors such as time, money and interest regularly drive the direction of inquiry and force a narrow but often effective approach. Yet we must wonder if a study of mental disorder can or should eliminate the individual. The ease of using an ECG to evaluate the heart does divorce the person from the organ, but need not take the position that the heart is identical to the person. We are more than our brains, and this acknowledgment does not diminish their role and function nor does it open the door to the premise that we are only our brains. The brain generates the mind, which is the seat of meaning, and its agent is the self. The study of mental illness must involve all three with no effort to reduce them to a single issue.

Dreams and dreaming serve as examples of the difference between an empirical study of a fixed phenomenon and an interpretation of a meaning. The ordinary dream book often attempts to make a one-to-one decoding, such as a black cat stands for, or a sword indicates, etc. Yet the analysis of dreams can be decoded or interpreted only by the free associations of the patient. Thus, there can be only personal or individual meanings to any single percept such as a black cat. And these individual meanings depend upon the particular life history of the person. To the degree that we all share experiences, we may conclude that we may well share meanings, but again that requires personal concurrence and cannot ever be automatically assumed. If one had grown up with a black kitten who was a constant childhood companion, then surely a dream of a black cat has a meaning for that person that distinguishes it from anything that approaches a universal meaning.

The inevitable question that arises in the effort to solve the conundrum of this triad is whether the elements can be studied individually. Of course we know that such studies are pursued, but we may be unwittingly making a mistake of premature categorization. One fascinating example of this is the study of mirror neurons, which has been eagerly embraced by those investigators who describe the need for and occurrence of "mirroring" behavior. In an investigation of animals watching other animals carry out a particular action, changes in the motor part of the brain mirrored those in the brains of the animals they were watching. Thus the brain activity mirrors that of the animal being watched. However, special neurons are not required for this and Victor Gallese, one of the discoverers of mirror neurons, refers to this imitation or action as "resonance behavior" (Gallese, 2008). Resonance is not confined to the brain but to the entire pattern of movements of the body. As Sheldrake notes, "watching other people engaged in sexual activity stimulates erotic arousal by a kind of resonance. The entire pornography industry depends on it" (Sheldrake, 2012, p. 204).

Materialism, dualism and self-organizing systems

The mind-brain problem has been a struggle for a variety of scientists and philosophers, ranging from Paul Churchland who claims that mental states are but "folk psychology" that will be ultimately replaced by explanations of nerve activity (Sheldrake, 2012, p. 112) to Galen Strawson (2006), who believes that atoms and molecules have a primitive kind of mentality or experience. The latter believes that self-organizing systems have a mental as well as a physical aspect. One can long puzzle how all those simple letters of the alphabet manage to join together to achieve a Shakespearean sonnet or how all those atoms and molecules combine to form a Picasso painting, but the supposed mystery that people such as Dennett insist is illusory (Sheldrake, 2012, p. 109) is readily solved by our opening the door to agency; i.e., the self. The self is the director and organizer of the mind. It is, of course, not to be found in the brain any more than beauty is to be found in a single place in a Shakespearean sonnet. The subjective world is not the objective world, but it is not therefore nonexistent. It is to be studied in its own right just as the brain and the mind are to be studied both individually and collectively.

We pay a price by reducing psychological phenomena to neurologic activity, just as we pay a price if we ignore the brain activity responsible for our subjective experience. The self is not the center of gravity as Dennett (2013b) suggests. Rather, we need to entertain different ideas such as those seen in holography to explain what Nöe says; i.e., "we are in the world and of it. We are patterns of active engagement with fluid boundaries" (2009). We need to open our minds to the Heideggerian idea of our being in the world rather than standing outside and perceiving the world.

In *Models of the Mind* (Gedo & Goldberg, 1973, p. 19), an effort was made to employ multiple models to explain and understand different developmental stages of the mind. A hierarchical overall model was introduced to encompass these varied viewpoints. A field theory of the mind goes beyond this perspective and suggests that the brain is not a storage system of memories, and thus the mind is not within the brain but rather is extended into the world. This will, no doubt, lead to a rethinking of transference and allow us to include both the patient and analyst thinking of each other when neither is present. Concepts such as projection and introjection, and especially the mystery of projective identification, may offer themselves to more cogent explanations in this concept of the extended mind. This work must be left for later. The present need is to delineate the mind and its study from that of the brain and the self. One is a vital organ composed of neurons, synapses and computer-like activity. One covers the vast area of meaning and offers us an entry into interpretive science, which stands apart from empirical science. And one is the seat of agency, which defines our individuality. It is necessary that the three are never reduced to the one or the other, despite the lure of reductionism.

No doubt a critic will claim that these are arbitrary definitions, but meanings are necessarily subject to a variety of interpretations. Thus the exercise in distinguishing between brain, mind and self is one of designing workable definitions.

Some solutions

Enactivism is a theoretical approach to understanding the mind. It emphasizes the way organisms organize themselves by interacting with their environment (Varela et al., 1991). In contrast to this viewpoint, psychoanalysts are familiar with those representational theories that state that the brain receives input from the world and builds up an internal picture or representation of the world. This internal picture is felt to be a more or less accurate rendition of the world and is changed by correcting one's inaccurate rendition. The externalized viewpoint makes the claim that accuracy is not the same for everyone inasmuch as our very involvement in and with the world creates the world. Of course, these are philosophical arguments, but they have become relevant to our efforts to differentiate the brain from the mind from the self.

If the brain is closed off from the world save by the input from the sensory organs, then, for whatever one may introduce from either a neurologic or psychologic cause, a place must be made for what this input means to us. Our involvement with the world leads to our seeing the world in a personal way, which inevitably leads to each of us constructing our own individual world. Thus the mind, wholly dependent upon and generated by the brain, is the arena wherein we see what the world means to us. Meaning becomes a larger concept than the brain. Despite the wish to reduce all thinking to brain activity, we find that the bulk of cognitive scientists somehow are forced to talk about the mind. In a fascinating discussion of predictive brains by Andy Clark (2013), the philosopher Daniel Dennett writes about the "cuteness of babies," which he situates in the mind of the observer and proceeds to explain as having evolved from our nervous system. Of course, this property is not able to be pinpointed in any particular area of the brain, but is to be relegated to the arena of "folk psychology," which, one would guess, will ultimately be eliminated in favor of more exact anatomical features. Until that time we had best stay with the mind. The claim that adapting a "neurocomputational perspective" (Churchland, 1989) will deliver a better understanding of agency than the "folk psychology" framework is really saying that it will be a different perspective employing a different language.

We cannot easily escape from the idea that we are agents that act upon the world and in so doing change the world. In what may be an unfair comparison, any effort to reduce a Shakespearean sonnet to its words will essentially miss the meaning of the sonnet. In a negative review of a book on brain imaging, the reviewer scoffs at the idea of treating the person rather than the brain by stating, "do they really think that dealing with the personal and emotional needs of addicts is dealing with something other than a disturbed brain?" (Gross, 2013). That same plaint could be expressed about, say, *Hamlet* in that the whole point of the play is surely in the words. Like it or not, people are surely their brains, but they are more than their brains as well. Indeed, different interpretations of *Hamlet* by different actors yield different meanings. However, isolating the brain from its environment alters the study of the mind.

We may do best in a recognition that the brain is in and with the world. Together they deliver meaning, and the agent of this organizing activity is the self. Although mental illness is a brain disorder, it is also a disorder of the person involved with the world, and that involvement is carried out by the self. One may study the brain in isolation, all the while recognizing its isolation, or else investigate the mind in meaning, all the while recognizing the multiplicity of meaning, or concentrate on the self, all the while recognizing its composition as including others. One possible yield of keeping these distinctions in mind might be an ability to determine exactly what treatment would work best for what form of pathology. Some patients do best with medication. Some need psychotherapy. Some require an analysis, and even there we need to determine what kind of an analysis. Last but not least, some patients require combinations of medication and therapy or analysis. We must move beyond the idea that these choices are personal preferences in a game of chance. Different illnesses require different interventions: one size does not fit all.

Indeed, if understanding is the essence of "being in the world" and if the way we achieve understanding is by way of an empathic connection to another, then an initial task is to better unpack that word "understanding" in order to aid in distinguishing between the three puzzles of brain, mind and self.

Chapter 2

On understanding "understanding"

How we understand the meaning of the word "understanding"

Psychoanalytic understanding in its status as a hermeneutic activity is different from other forms of understanding. It is distinct, for instance, from psychodynamic psychotherapy, which may be able to establish itself as an empirical science. Empirical science deals with rules and establishes facts, while hermeneutic science deals with meanings, which cover a wider area than facts. Meanings offer a different kind of explanation than facts offer. Psychodynamic psychotherapy can be differentiated from psychoanalysis on the basis of their being different forms of scientific activity. Psychodynamic psychotherapy is not a diminished or lesser form of psychoanalysis inasmuch as it employs both interpretation and other therapeutic activities, whereas psychoanalysis is best seen as restricted to interpretation. The understanding that results from psychoanalysis is unique to psychoanalysis.

My intent in this chapter is to offer a distinctive delineation of psychoanalytic understanding as it stands apart from other forms of understanding. A secondary goal is to better situate the status of psychodynamic psychotherapy. We are all familiar with the relative senses in which we invoke the concept of understanding. For example, we may know and understand that an automobile needs gasoline to function, but we may not know exactly how the gasoline is used. We may know that the gasoline is somehow acted on in the carburetor, but we may be a bit vague as to the exact workings. Understanding or know-how, even if primarily cognitive, has levels and branches and so must be recognized as being employed differently in areas that can be empirically studied and measured, as in my gasoline example, compared to areas considered to be interpretive or hermeneutic studies, as in the humanities and psychoanalysis. We may be able to effectively compare different brands of gasoline and different engines, and thereby choose among them, but may not be able to do so with different interpretive efforts. So, too, we should not equate some forms of understanding with translation. We may be able to translate Chinese into English and from that effort know what a word or sentence means. That too is a form of understanding that differs from, say, an interpretation of a Shakespearean sonnet or play. Words, sentences and paragraphs have meanings, but some are also open to interpretation, which in turn opens them to varieties of meanings. *Hamlet* in any language can be interpreted in different ways and so is open to be understood in different ways. A translation can make things understandable but may in turn be open to a number of interpretations.

Essentially we must distinguish the wide area of meaning (i.e., what things mean to us) from the area of facts and truth (i.e., what can be measured or explained in a manner considered scientific). Meanings are open to a number of interpretations, all of which may be true, and so may allow us to pick and choose. Facts and truths do not allow for interpretation and choice, and they are but one category of meaning. Thus, meaning covers a large area.

The task for our efforts is to distinguish between what we aim to do when we (1) understand something, (2) explain something and (3) translate something in order to make it more understandable. We shall initially restrict ourselves to the two categories of meaning and explanation. The first, hermeneutics, is an open category that depends on individual comprehension. Thus, a passage in the Bible or of a Shakespearean play may well mean different things to different people. It is necessarily subjective. However, in contrast to this is the category in which we can be factual and objective. We reserve that for explanation in empirical science. That is where facts that are beyond individual opinion should emerge and so allow an end to the pursuit of truth regarding a particular question. Translation makes for a subcategory. In a review of a translation of *The Divine Comedy*, Luzzi notes the struggles of translators who need to preserve the meaning of a work by "liberat[ing] the language imprisoned in the work" (Walter Benjamin) while making it more understandable (2013, p. 13). The translator thus strives for a more understandable product and hopes to keep the text's basic meaning intact. Something like this often occurs in our psychoanalytic conferences, especially when divergent theoretical perspectives are being offered. The ego psychologist attempts to "make sense" of what the Kohutian analyst is saying by translating the latter's comments into a language that is more congenial and compatible. As in many translations, something of the original intent may be lost, but the ego psychologist is truly "liberating" the language imprisoned in the jargon of the Kohutian. For the Kohutian, however, the meaning is lost.

As long as we consider psychoanalysis to be a pursuit in search of meaning, it is likely that the present state of pluralism may merely reflect the unavoidable possibility that every analysis will be different, composed as it is of a narrative constructed by two individuals, each with a different vocabulary. This should not mean that the number of meanings is inexhaustible, but rather that the hope for a single, accurate meaning is unlikely. Each analytic theory strives to be a sufficient explanation, but, much like the many excellent interpretations of a Shakespearean play, they are satisfactory to some and lacking for others. Unfortunately, for those who might yearn for objectivity in psychoanalysis, this is seen as a flaw. Psychodynamic psychotherapy is often an attempt to achieve that objectivity. It often does so by focusing on an aim or goal, usually involving a circumscribed symptom. If a patient is depressed and a therapy relieves the depression, then the process may be described and replicated and tested. We may then say that we both understand and explain the depression. If that same patient undergoes an analysis, the result may be the same in terms of symptom relief, but a multitude of meanings of that depression may be developed with an accompanying challenge as to what exactly was responsible for the symptom relief. Indeed, the analysis may go beyond what is needed for that limited goal and may even be unsuccessful in that larger pursuit. Inasmuch as our different psychoanalytic theories are directed at different psychological issues, it may well be that some are more effective than others in providing symptom relief and yielding explanatory meanings. Where psychotherapy seeks to arrive at symptom relief, analysis seeks to arrive at a meaning that best reflects the theory that generates the procedures. Debates about the efficacy of one analytic theory over another are much like choosing one interpretation of *Hamlet* over another; essentially it depends on fealty to a communal set of beliefs. By contrast, debates about one psychotherapeutic approach over another are resolved by symptom relief. We regularly hear of cases that seem more directed toward clinical improvement, with minimal concern for understanding meaning. The hope that psychotherapy can become more scientific may be realized. The same hope for psychoanalysis rests on a misunderstanding of its hermeneutic nature.

The word

The title of this chapter is something of a challenge because the word "understanding" is called on, when we aim to understand the concept, to function both as a verb (the action of understanding) and as a noun (the state of understanding). Philosophers have devoted a good deal of time and energy to this project, but rather than giving an extensive review, I can simplify the issue by using Brandom's distinction between "algebraic" understanding, which follows rules, and hermeneutic understanding, which is more basic and involves concepts and meanings that can never be systematized (2008, pp. 212–216). Ordinary discourse or conversation is regularly an effort at grasping meanings and so often becomes a hermeneutic exercise in the attempt to interpret what another person has said. However, when we are in the gas station the signs are usually clear as to the price of the gas and how to pay for it. Usually no further interpretation is necessary. Although some may hope that all human discourse and forms of writing may come to admit of equally clear and exact understanding, the richness of language makes rule-following impossible. We are regularly confronted with a multiplicity of meanings, in our conversations and in a variety of artistic pursuits. Psychoanalysis should be recognized as a hermeneutic activity; efforts to make it an empirical science on par with brain studies or to reduce it to some collection of core beliefs (Tuckett et al., 2008) are, by definition, acts of futility.

Though one may insist on making a clear distinction between extrospection (the observation of external phenomena) and introspection (the observation of internal states of mind), this distinction is apropos to neither hermeneutic nor empirical studies. Holzman (1985) seems to insist on this difference in order to condemn hermeneutics, but that such a distinction in fact holds is merely his opinion. If a salesperson wishes to sell something to a customer, he or she regularly and almost automatically aims to determine what the product means to the customer by a combination of empathic and empirical data gathering. This combination consists of the price of the product, the empathically grasped needs and interests of the customer, the needs and interests of the salesperson (who may be on salary or working on commission) and so on. True facts are ever constituents of a hermeneutic activity, and we reach meanings by incorporating these truths and their impact on meaning. When a scientist arrives at a conclusion based on empirical findings, he or she almost always adds in the potential meaning of the discovery in areas ranging from personal importance to the long- and short-term impacts of the conclusion on other investigators' work. Ideally one may wish to eliminate certain emotional factors in some empirical studies, just as one controls for a number of possibly interfering factors in running any experiment.

This potential interference may be responsible for the erroneous assumption that hermeneutic studies stand in opposition to empirical studies. It is probably better to consider the two types as relevant at various times and to various degrees, inasmuch as it is a rare fact that has no meaning whatsoever. Thus, psychoanalysis as a hermeneutic activity always incorporates certain truths and facts in its work.

An analytic exercise

A patient in analysis tells his analyst about his girlfriend, whom he loves and plans to marry and with whom he feels a happiness and contentment he has never before felt with a woman. Save for one thing. He so wants her to tell him that he is the best lover she has ever known. She has been married before and is quick to assure him that she loves him and is happier with him than with anyone she has ever known. But she never tells him that he is the best lover she has ever known, and he cannot seem to eradicate the wish that she say so from his mind. So patient and analyst try to understand the meaning of this persistent thought, which has seemingly risen to the level of a symptom. He cannot seem to stop thinking of what he so longs for, and that seems forever to elude him.

The patient feels he understands the origin of his thought, since he has always wanted to be number one in everything he has done, and this is surely no different. His analyst notes that this conforms to the patient's grandiose fantasy of himself, along with his singular position in his family of origin. The patient confirms that he fears the embarrassment and humiliation of weakness and inadequacy, a fear that seems to be at the heart of this preoccupation with greatness. This analytic explanation seems to suffice, but seems also to lack certain crucial elements. Is his problem merely one of narcissistic preoccupation because of something like a failure in early mirroring, or is it either more than that or other than that? We begin this journey amid the multiplicity of meaning. The patient says that his mother was a model of discontent, and that only he could make her happy. He insists he is treating the relationship with his prospective wife much like a performance designed to elicit praise from his mother. This too seems a perfectly adequate and even comprehensive analytic explanation. But yet, it does not seem quite complete. The joining of the two interpretations is good, but more seems needed. We are in the grip of these many levels and branches of interpretation, which do not neatly add up to a sense of completeness and reasonableness, a hoped-for culmination and finality.

At this point in the analysis (which, by the way, is not presented as representative of good technique or any particular theoretical program), the patient wished to go no further; though his obsessive thought continued unchanged, he felt satisfied with these explanations. He wondered, though, whether the analyst, now armed with these psychoanalytic interpretations, could tell him something that would put him at ease with this troubling inner turmoil. The analyst, relying on a general rule that the persistence of symptoms may mean that an interpretation is faulty, and feeling that more information was necessary, turned to the transference and the patient's hope that his turmoil might be eased. This serves to illustrate a problem inherent in the multiplicity of meaning: there often is not, and cannot be, a definitive moment of ending. In a recent psychiatric presentation (which, to save embarrassment, will not be referenced), the clinician, in reporting a similar narcissistic issue in a patient, regarded his description and its explanation in DSM terms sufficient to explain the problem. Psychoanalysis, by contrast, allows the coexistence of multiple explanations with no clear endpoint or resolution. This question of knowing when and where to stop is a rather common, perhaps necessary, aspect of hermeneutic inquiry. It is often resolved in a treatment with an easing or cessation of symptoms, along with the fulfillment of a particular theoretical vision. If, for instance, one believes in the universality of the Oedipus complex, then its conscious explication and elaboration is the goal and so exists as an endpoint. Of course, one can reach such an endpoint while still maintaining a belief that more remains to be done. Analysis is capable of coming to points of resolution while still being open to further inquiry.

When our patient asked his analyst to do what the analyst felt was beyond his ability, the analyst asked for more information about the patient's father and his relation to the mother's deficit in satisfaction, and about the father's status in the family. The father had always been described as a well-meaning but weak and passive person. Very little, outside of going to work and an occasional visit to the ballpark, was expected of him. The patient insisted that the father was but another link in the chain of disappointments and discontents that his mother had chronicled throughout his life. He himself was certainly not like his father; he could even remember how his mother had made this very point. The analyst felt that the patient's request for a solution to his problem, coupled with the inevitable failure to provide one, was a reliable transference confirmation of his now altered theoretical explanation. His understanding had been modified while retaining its fundamental status.

Understanding does not often lend itself to neat solutions. This is so because of the many meanings it tries to incorporate, as well as their possible contradictions. Contradictions appeared to the analyst in the form of his patient's reliance on and expectations of the analyst, who felt admired by and respected by the patient. At one point the patient wanted to increase his session frequency, only to be stymied by the financial hurdle that would pose. Yet there was something rather contradictory about the patient, who, while seemingly dependent on the analyst, was at times dismissive of him and just as often wanted to cut down on sessions.

One of the common problems that emerge in an analytic understanding of a patient is the continual modification of the narrative being constructed to render what would be a complete explanation. Some points may be stressed while others are ignored or minimized. The story is often twisted and massaged in order to offer a complete and integrated explanation, quite apart from the lack of satisfactory verification. As the treatment continued, the patient began to talk more about his preoccupation with the men who may have been involved with the woman he was trying so hard to please. This opened up a new set of associations about his jealousy, which now began to take center stage and appeared more intense than his initial symptom of wishing to be the best lover. No doubt psychoanalysts may feel that this is a new and very rich area for exploration; this opening of new possibilities is a common and expectable issue in the interpretive process. Empirical science aims for tried and true resolutions, while hermeneutics may appear to be unending.

The analyst proceeded to modify his basic interpretation of the patient's wish for his girlfriend to say he was the greatest lover in the world. Added to grandiosity and narcissism as explanations, along with the wish to be his mother's favorite, was his thwarted identification with his beloved yet castrated father. Now the analyst began to worry. Was there no place for other theoretical ideas, ranging from the depressive position to splitting to desire?

Indeed the patient did improve with the interpretation about the father, but lots of patients get better for all sorts of reasons. There seemed to be no tight fit here between understanding and improvement, although even that conclusion was suspect, because at times the connection was quite dramatic. The only feasible conclusion was that understanding is something that makes sense and seems to bring together most of the questions to be answered. It offers not the neatness and finality that is so appealing in the empirical sciences (e.g. the patient either has pneumonia or he does not) but instead only a high probability of being right. Indeed, being definitively right or wrong began to fade as a reachable finality here in favor of the attitude "that is as good an explanation as any." The possibility of other, better explanations shifted from being itself a problem to the previously untenable position of being a virtue.

Thus far we have concentrated on how the process of understanding functions in most conducts of analysis. We now turn to how the process in itself impacts the analysis or psychotherapy.

Being understood and feeling understood

I will now distinguish between two phenomena. One is seen with a person who feels understood; it is perhaps best seen when a patient feels that someone else, in this case an analyst or therapist, understands him or her. The other is found with someone who has succeeded in making him/herself understood by someone else, be it a teacher with a pupil or a therapist with a patient. A subsidiary effort will be to distinguish psychodynamic psychotherapy and its place in science.

Ronald has long suffered from what has been diagnosed as chronic depression, and he has endured a series of treatments ranging from antidepressant medications, transmagnetic stimulation, electroshock therapy, to a prolonged hospitalization with dialectical behavioral therapy. He has now enlisted in a psychoanalytically oriented psychotherapy, but with little hope or enthusiasm. He describes a stable marriage, though it has its disappointments. His wife is totally devoted to Ronald and to a host of other people whom she serves as teacher and supporter. She wakes at 5:30 every morning to read her Bible, and when Ronald confessed his sexual infidelity to her, she was quick to forgive and never mentioned it again.

While earlier we subjected the analytic process to a study based on hermeneutics, we now reverse the investigation by analyzing the effects of hermeneutic procedures. Thus, when the therapist suggested to Ronald that he might find it difficult to live with a woman who, by all accounts, is something of a saint, Ronald agreed. When it was further suggested that perhaps he had a good deal of pent-up anger he was unable to express, Ronald smiled and agreed. He felt understood.

Ronald is presented as an example of how being understood leads to an increased feeling of well-being, and that this can be explained from a variety of psychoanalytic perspectives. This may lead to a category of understanding that is solely devoted to therapy. Because empiricism is dependent on hermeneutics, an empirical psychotherapy may arise from our analytic understanding, a therapy more rule-bound and factual than psychoanalysis itself. We can explain Ronald's improvement as the result of something preconscious or unconscious coming to conscious awareness, or as the consequence of a feeling of connection in a needed relationship, or offer a host of other explanations. Just as a dynamic, analytic orientation is needed to make sense of how being understood can be ameliorative, we earlier saw how understanding as a process can explain the workings of psychoanalysis. Although different theoretical stances can explain this improvement in varied and perhaps even contradictory ways, one cannot conclude that there is a single correct answer. Despite the pressure to elaborate a rule-guided, perhaps ultimately measurable procedure to serve as an answer, it seems more likely that there exist a number of reasonable explanations. At the present state of our knowledge, with all its pluralism, it may well be the case that a number of different ways to explain Ronald's improvement may be correct. Proponents of each explanation may feel theirs is correct and best, and no single champion will emerge.

To be sure, it is not the case that any and every explanation will allow Ronald to feel understood, and so surely there is some necessary component in its achievement. Here too we enter a complex and elaborate philosophical thicket, which does, however, enable us to resolve this dilemma. In brief, today's understanding of pragmatism as "what seems to work" allows for membership in the community of psychoanalytic explanations that allow Ronald to feel understood (Goldberg, 2002). The therapist who interpreted Ronald's rage at him enjoyed a complementary good feeling at seeing his idea be productive and accepted.

But not every such intervention yields such a positive result, and so it becomes necessary to examine and explain the impact of both our successes and our failures. We often turn to the concept of resistance to handle the latter, and that idea serves as an opening to explain its companion effect: efficacy. We feel effective when our interventions work and seem to support our belief system. Our membership in a community of like-minded therapists involves us in claims we feel to be true, and the verification of these truths is needed to allow us to prevail. When we are effective, we feel justified in our beliefs, which in turn bolsters our identity and our allegiances. That a number of systems of belief can be equally effective is further testimony to the fact that there need be no single set of truths that enables us to achieve understanding. Lots of psychotherapies work.

Empiricism and neuroscience

Beset as we are with a multitude of meaning to explain what we do, as well as how we react to what we do, it is of some interest for us to confront the demand that we be more exact and open to empirical validation. Hermeneutic understanding should not be, but often is, placed in opposition to empirical studies dealing with facts, truths and predictions. Recent advances in neuroscience have led to the hope that psychoanalysis can ultimately become more fact based and pared down to a single system of core beliefs. If this does not happen, then perhaps psychoanalysis is doomed to disappear. Or perhaps a new version of analysis can meet the demands of empiricism.

One can imagine that an MRI of our angry patient Ronald would reveal that some aspects of his brain that correlate to anger are not properly discharging. Either telling Ronald that he is angry based on his MRI or causing these anger centers to discharge might do the same work as the constructed narrative presented above. We do know, however, that the understanding we aim for is not simply a cognitive or conscious construct but rather is a more holistic comprehension. Efforts to bring understanding to a more empirical basis are essentially efforts to reduce a complex system to a simpler one, and this is often unsuccessful. I do not mean to dismiss the importance of neuroscience or of brain studies, but rather to place them alongside psychoanalysis, instead of viewing them as its replacement. Each has a place, and these places are simply not interchangeable. Nor should one hope they ever will be. Explaining the meter of a Robert Frost poem does not help us fully understand its meaning, as much as the meter might contribute to that meaning.

... and psychotherapy

If one takes the stand that psychodynamic psychotherapy is a subsidiary form of psychoanalysis, it can be seen as on par with the multitude of nonanalytic psychotherapies, some of which are amenable to measurement and procedural instruction. These therapies are effective for a variety of reasons and should be accepted for this efficacy rather than for any measure of fidelity to psychoanalysis. Although analytic psychotherapies are dependent on the psychoanalytic theories from which they derive, they differ from psychoanalysis in that they often follow rules and aim primarily, if not exclusively, for symptom relief. In contrast, what the psychoanalytic process aims to achieve is a meaning and understanding that is felt to be complete and thorough. If symptom relief also occurs, it must be as part of the meaning rather than as the goal of treatment. In Ronald's case, enabling him to get in touch with and experience his anger was a rule-guided operation that resulted both in a certain level of understanding and in symptom relief. Ronald came to feel understood, through a rule-guided process seen in many forms of treatment. These processes are open to empirical study and are subsumed under an overall hermeneutic study that remains open-ended. Essentially, it is unending. However, other aspects of meaning are not subject to empirical study, and do not relieve symptoms. But they do lead to increased comprehension and understanding.

Connectedness and development

The centrality of empathy in enabling one person to understand another is fairly well accepted as the indispensable initiator of understanding. However, empathy is often no more than the beginning of a long journey in the construction of a narrative of someone's life. The accompanying salutary effect of being understood cannot itself fulfill the complex demands of the analytic process, which is in one theoretical scenario a series of empathic disruptions and unions. It is an error to consider connectedness sufficient. It is best thought of as only a necessary start.

A similar error occurs when a patient's development is seen as incomplete or aberrant. Analysis surely must be directed at allowing normal development to proceed, but that should not be seen as the arena for hoped-for new growth. Franz Alexander's adventure in the "corrective emotional experience" (1961, p. 329) stands as a fundamental warning against psychoanalysis serving as "replacement therapy." Kohut was adamant in stating that analysis should do no more than allow the patient to pursue a development that has been thwarted. But that was to be accomplished after the analysis is over. In our zeal to be therapeutic, we often feel we must offer patients what was lacking in their lives. We do best, however, when we simply afford them the opportunity to grow, rather than offering ourselves as replacements for an imagined deficit. Limiting our efforts to attaining and maintaining connectedness, as well as acting as carriers of developmental needs, is misdirected. Psychoanalysis is an activity of interpretation rather than a primary modality of treatment.

Discussion

Psychoanalysis can never join the ranks of empirical science, which measures and compares procedures and results. In such a pursuit, patients with identical or similar pathology can be treated with a variety of modalities in order to determine which is the most effective and least costly. The never-ending hope that analysis can take its place in this endeavor must be abandoned. No doubt many forms of both psychodynamic and nonanalytic psychotherapy can enlist in this research activity as they become more rule-bound and systematized. Cognitive behavioral therapy is a good example of a psychotherapy that has a manual of rules and procedures and that lends itself to comparative evaluation against both medication and other active interventions. The primary reason for this supposed limitation of psychoanalysis is that psychoanalysis is a hermeneutic science that seeks to understand and discover the meanings of a person's malady. Because there are no "true" meanings to be unearthed, the process of understanding will allow a variety of meanings to emerge. The same, of course, can be said of works of art. The Bible, the works of Shakespeare, the poems of Robert Frost and the plays of Samuel Beckett all mean different things to different people, and there is no end to efforts to explain them to the willing listener. And most of these explanations are both good and comparable, though they run the risk of interfering with the pleasures of understanding. Hermeneutic science is not a lesser form of science; it is just different.

Explaining understanding

The old argument over whether psychoanalysis is an understanding psychology or an explaining psychology was said to be effectively answered by claiming that we explain what we first understand; it is a two-step process (Kohut, 1984). Unfortunately, all our explanations are not on the same footing, and this is often true as well of what and how we understand. Hans-Georg Gadamer, often considered the leading spokesperson and populizer of hermeneutics, has answered this problem with his claim that narratives or texts are resistant to arbitrary meanings and so can question some interpretations. Likewise, he reminds us that the meanings that we seek arise in and during the process of understanding; nothing has a meaning "in itself" until subjected to understanding (Dostal, 2002). The present-day pluralism of psychoanalysis demonstrates how a number of theories are capable of both reaching different meanings and explaining the same meaning in different ways. The theories employed in explaining are, for the most part, rule-based and conformable to "algebraic" rather than hermeneutic understanding. Thus, ego psychologists, Kohutians or Kleinian or Lacanian analysts can *explain* a patient according to the particular rule book by which they operate. Indeed, they may also *understand* the patient either in a similar manner or in a different one. There is no true meaning, but quite often there is a remarkable consensus of understanding that is shattered only when a particular explanatory theory is employed.

Perhaps if the scientific status of psychoanalysis were better recognized and appreciated there might be a diminution in the competition between theories and a redirection of interest away from considering the primary purpose of analysis as treatment. Holzman (1985) was prophetic in this warning, but a compromise may be in the offing. That is, psychoanalysis is an adventure in understanding oneself with a possible therapeutic effect. Psychodynamic psychotherapy is a treatment for mental illness with the possible side effect of understanding. In this manner the repositioning of psychodynamic psychotherapy as a legitimate rival to other forms of treatment could be realized, and psychotherapy could be studied not as a test of fidelity to psychoanalysis but more as a bona fide therapeutic activity. Unfortunately, much psychotherapy is taught as a watered-down version of whatever analytic theory spawned it, rather than as an independent agent of alteration. Every now and then, when a psychotherapy seems to have failed, an analysis is suggested, as if one were to apply a more potent version of a similar product. And every now and then, when an analysis fails, it is judged to have done so primarily because it has brought symptom relief rather than an expansion of understanding. Perhaps Gitelson was correct to have insisted that symptom relief be considered as side effect of analysis (Goldberg, 2001b). As difficult to comprehend as it may seem, Freud as psychiatrist misdirected his discovery in his pursuit of analysis as treatment. We would do well to restrict psychoanalysis to its proper place as a hermeneutic science, and to see whether we can develop psychodynamic psychotherapy as a therapeutic tool in its own right.

Concluding remarks

Philosophers of science, led by many of the great names in philosophy but best represented by Gadamer, have shown science to be divided into empirical and hermeneutic categories. Psychoanalysis as an understanding psychology can be approached with a number of theories, each of which can lead to a particular form of understanding that may at times result in the removal or diminution of the signs and symptoms of mental illness. Just as patients may benefit from analysis, so too do the practitioners of understanding benefit from its application. However, hermeneutic science cannot be a rival to empirical science in establishing rules of activity and forms of measurement. Rather, it may be profitable to carve out a separate category of psychodynamic psychotherapy as an empirical, measurable scientific activity. Psychoanalysis, however, must remain a hermeneutic science devoted primarily to understanding, not to the care of mental illness. We should encourage the establishment of psychodynamic psychotherapy as a proper rival to other treatments for mental disorders. We should also recognize its difference, which surely makes a difference.

Chapter 3

On the scientific status of psychoanalysis and psychotherapy

As a student in psychoanalysis, an untraceable quote attributed to Lawrence Kubie always stood out in my mind. Kubie was purported to have said that at the end of a well-conducted analysis the analysand would forget everything about the analyst, including his or her name. The explanation behind this statement was that the analyst was the primary seat of all transference and thus, when all of the many transferences were interpreted and thereby dislodged or dissipated, there would be nothing psychologically left of the analyst. That is to say that he or she would no longer be necessary. Inasmuch as transference is said to be either "new editions or facsimiles" or "revised editions" (Freud, 1905b), the analysis would allow these to be made conscious and so removed or destroyed. What might remain of the analyst might be considered to be the real attributes of that person that may be thought of as mere add-ons or features that may or may not facilitate the work of analysis, but ultimately must be dispensed with as well.

If one visits a restaurant in order to eat, it may be necessary to also frequent the restroom of that establishment. Yet the visit to the restroom is to be thought of as a necessary but secondary reason to attend that restaurant. Indeed, if a supposed patron comes in and uses the restroom and leaves without ordering and eating, then that is felt to be an abuse of the establishment and so may be seen as an act of either misbehavior or confusion. This is much like the analyst or analysand who claims a benefit from an analysis that seems to some to be not exactly what the analysis is supposed to do.

I recently received a letter from a patient whom I had seen some time ago, and one whom I could not clearly remember. She wrote that she was not sure what she had gotten from her treatment, but that the most important thing for her was the feeling that I cared. Now I am sure that that feeling could itself be a proper focus for a transference investigation, but it did allow me to wonder about a number of features of analysis that, although central to any and all analyses, could be seen as secondary add-ons. To be sure, what is remembered may not be the crucial ingredient; however, some add-ons often attain a certain prominence that make them seem central. One factor that seemed applicable to these necessary, but essentially secondary, aspects of analysis was that of satisfying Kubie's criteria of vanishing. Could my ex-patient achieve the very necessary feeling of being cared for in areas of her life outside of this long past event? In other words, might she have no longer remembered me in that particular manner?

Surely there are a number of features of an analysis that are both central and crucial to its conduct but may not qualify as unique to analysis. Issues such as regularity may become singularly important in some analyses. The centrality of listening has been often cited as the crucial component in some analyses (Schwaber, 1986), and indeed some analytic patients may claim "being listened to" as the most memorable feature of their treatment. Listening and attention regularly connect to "feeling cared for," and the memory of my old patient joins here in the host of qualities and factors that both characterize psychoanalysis and yet seem shared to some extent by a myriad of what might be thought of as nonanalytic procedures.

A crucial voice in the debate over recent additions and emendations to psychoanalytic theory and clinical practice is that of Stephen Mitchell, who offers a distinction between the wishes of a patient and the needs of a patient (Mitchell, 1991). Although Mitchell also offers an extensive review in order to clarify the status of a wish as opposed to a need, he seems to be unable to make a definitive difference clear. Needs are called "ego needs" and should be gratified in those patients who are suffering from deficits and developmental arrests. This is true, according to Mitchell, of the more disturbed patients. The gratification of instinctual wishes may, however, be unwise in other patients.

Although earlier in his paper Mitchell offers a distinction between positivism and hermeneutics, he does not utilize this distinction in his effort to allow the analyst to further clarify the need-wish dilemma that he hopes to explain. Indeed, this dilemma has surfaced before in psychoanalysis in the form of a debate as to whether analysis is an understanding psychology or an explaining psychology. The claim that it is both often leads to more confusion than enlightenment, yet is essentially the answer to Mitchell's struggle over "gratification," which he resolves with a new "relationist" perspective.

Our language and its users allow for a wide degree of flexibility and nuance, and there is no doubt that wishes and needs are often used interchangeably. It may be safe to say that wishes are primarily psychological while needs are perhaps best seen as biological. Indeed, Mitchell suggests as much in his indication that some patients who suffer from developmental arrests need some forms of activity (i.e., more gratification) to allow development to proceed. Kohut made clear that the recognition of certain selfobject needs should be allowed to be gratified outside of the analysis, and so he distinguished between the initial psychological understanding and the subsequent fulfillment of these needs (Kohut, 1984). It may be helpful to see the first as the hermeneutic activity of psychoanalysis and the second as a fulfillment of a biological program based upon the empirical, developmental set of standards and goals. We understand wishes. We explain normal and abnormal development, and therapy may be utilized to correct the abnormality.

Obviously many psychoanalytic efforts mix the activities of understanding and explanation. Kubie entertained a somewhat idealized version of psychoanalysis in his restriction of it to interpretation of transference. However, that restriction may serve to highlight the fact that interpretation, or the hermeneutic activity of psychoanalysis, is different from any and all forms of need gratification that may well be pursued in order to follow a preferred developmental theory. This latter is a product of the norms that our studies have formed and followed over the years. We all have a set of rules or programs that we consider applicable to our patients; i.e., they should behave, think and feel in certain acceptable and uniform ways. Failures to do so regularly lead us to become involved in any and all sorts of "gratification" that can be seen as therapeutic and thus growth enhancing: our ideas about normality serve to guide our theory of treatment.

Empirical science seeks to discover categories that are widely applicable and so to derive standards and goals that are considered to be normal or abnormal. Blood pressure can be measured and determined to be "within normal limits" or else to be treated. So, too, are many psychiatric categories carefully monitored and descriptively categorized as normal or pathological. Such normative considerations are not usually brought to bear on interpretive science, which aims to refrain from a search for correct or true decisions. There is a fitting quote from Gadamer that states, "in understanding we are drawn into an event of truth and arrive, as it were, too late if we want to know what we ought to believe" (Ingram, 1985, p. 49). One way to unpack the meaning of that quote – i.e., to interpret it – is to read it as part of a process that continually reveals what may be thought of as true. Hermeneutics involves a to-and-fro dialogue of questions and answers that continue until an agreement is reached. It does not attain universal truths (Ingram, 1985).

Case report

G. was a young man in analysis whose early life was best described as chaotic and unpredictable. He took readily to coming to analysis four times per week and insisted that he found it extremely helpful. This declaration puzzled his analyst, who felt that very little of what he had to say seemed to have much of an impact on G., whereas the regularity of the sessions appeared without a doubt to be of enormous significance to G. After his apartment was broken into, G. was forced to move, and the analysis proceeded to be consumed by the myriad of efforts and plans to constitute the newly formed place of residence. G.'s analyst recognized that G. needed a reliable and predictable segment of his life to serve as a sort of structure for him, a structure that had been sorely missed during his development. Thus a clear "need" was recognized and offered by G.'s analyst, who was asked to do little to allow this need to be realized. Once this was recognized and interpreted, G. was able to connect what he saw to be a deficiency of his to earlier events in his childhood.

G.'s analyst puzzled over the question of whether the relationship to the analyst was therapeutic in and of itself or whether the interpretations offered by the analyst were the crucial ingredient. His puzzlement was resolved when G. understood what he had derived from his analysis and was able to utilize other forms of regularity and predictability in his life. He indeed had been given an explanation of what he understood, but only the understanding per se was able to offer him the freedom to construct a structured life outside of his analysis.

As G.'s apartment took shape and began to serve as the stabilizing structure that was so lacking in his early development and that also took form in his analysis, he was more and more able to utilize his analysis for the interpretive work that it offered and less for the need gratification that he had so desired. One could posit that had G. been a different patient less in need of such a stabilizing structure, then the analytic work could have been solely directed to the ideal state so sought by Kubie. Perhaps G. had become that different patient who now lived in a different world.

Discussion

Some time ago, Thomas Kuhn upset the world of science in his book, *The Structure of Scientific Revolution*, with the claim that a paradigm change in solving scientific problems essentially had the scientist living and working in a different world (Kuhn, 1970). In a later paper Kuhn concluded that his work involving and explaining the "natural sciences" had to do with solving puzzles aimed at improving and extending the match between theory and experiment. What Kuhn terms the "social or human sciences" are interpretive and not directed at discovering the laws that govern behavior (Kuhn, 1991, p. 23). Kuhn struggled over the question of whether the human or social sciences would be directed to the sort of puzzle solving that the natural sciences would be involved with, but he concluded that the human or social sciences were "interpretive through and through." If a new paradigm does arise in the natural sciences it is not the product of a search, but rather comes about in an "involuntary manner and is often not recognized as such" (p. 23). However, the social sciences are devoted to interpretation and so "new and deeper interpretations are the object of the games" (p. 23).

If we apply Kuhn's conclusions to psychoanalysis we see that many of its practitioners work to make their interpretive interventions serve the goals of the natural sciences; i.e., we yearn for truths, and we follow the efforts of fitting our efforts to our theories. Yet the multitude of psychoanalytic theories reflect a lack of a single paradigm, and the multitude of psychoanalytic interpretations often invite differing goals. Indeed, within any particular theoretical approach, be it Kleinian or Lacanian, there is often no single, true interpretation that is employed and agreed upon.

Back to Mr. G.

After his analyst interpreted to G. that he sought the regularity and stability that had been so lacking in his early life, G. did indeed inhabit a different world, a world that he proceeded to perceive in terms of his new insight. Without that insight, he might well have found comfort in the organized structure of his newly furnished apartment, but he would be unable to be himself a source of that comfort. Effective interpretations alter the meanings that we attribute to the contents of the world in which we live. Essentially effective interpretations are designed not to gratify needs but rather to enable us to see things differently.

Differentiating psychotherapy and psychoanalysis

There can be little argument as to whether psychoanalysis is ordinarily meant to be therapeutic nor can there be any argument that psychotherapy is meant to treat and perhaps cure psychopathology. Psychiatry as the empirical science behind psychotherapy describes and classifies what is considered to be normal behavior and does so by a variety of methods, ranging from medication to all manner of psychotherapy aims to correct the abnormal. When Maxwell Gitelson (Goldberg, 2001b) declared that symptom relief was a byproduct of analysis, he likened it to the added-on restrooms of some dining establishment; i.e., it was not to be considered the goal of the endeavor.

If we liken the hermeneutic activity of psychoanalysis to biblical hermeneutics, we can agree that different readers derive different meanings from selected passages in the Bible, although some may insist that all the readings are religious. Yet some scholars read the Bible for its historic significance and some even for its literary impact. The shotgun marriage of psychoanalysis and psychiatry has led to a misalliance of the goals of the two. Whereas psychiatry and psychotherapy aim to realign behavior to an accepted norm, psychoanalysis aims to understand the meanings of behavior. If that pursuit leads to a "correction" of deviant or incorrect behavior that is all to the good, but it need not be the goal of psychoanalysis. Just as some people may get "religion" from reading the Bible, it need not be the exclusive goal of all such readings. We must also recognize that Catholics and Unitarians see quite different meanings in scripture, just as Kleinians and Lacanians see different meanings in a dream. One inescapable conclusion is that there is no true meaning in a dream, as much as proponents of different theories may claim there to be.

Efforts to compare different psychoanalytic theories and techniques may assume a commonality, such as the belief that Oedipal conflicts are at the heart of all symptomatology, that simply does not exist. Thus parallel efforts to transcend doctrinal differences are best seen as political in nature rather than "fruitful in the search for unanimity" (Tuckett et al., 2008).

Summary

Imagine a horizontal line, one end of which represents Lawrence Kubie's conception of psychoanalysis devoted totally to transference interpretations while the other end is devoted solely to therapeutic relations that gratify needs for growth and development otherwise not achieved in the life of a patient. Positions along this line represent mixtures of these two endpoints, with the possible hope of moving any given patient from a standing point close to need gratification to a point that represents something close to pure interpretation. Kubie and Gitelson represent an endpoint devoted to psychoanalysis seen as an understanding psychology and so devoted to interpreting meanings, while Mitchell is a more pragmatic therapist who evaluates a patient's developmental needs in order to allow further development to proceed to some point considered to be normative.

Inasmuch as our line is an imaginary one, it is important that one not bring a moral component to its examination. Empirical studies such as neuroscience are not better than hermeneutic studies such as psychoanalysis. They are just different, although they may well arrive at similar endpoints of increased well-being. Confusion results when one set of principles attempts to act like the other; i.e., when one end of our imaginary line masquerades as its opposite. Interpretation does little for developmental deficits while need gratification by way of a relationship may offer little to alter one's understanding of oneself and others. Unfortunately, our present diagnostic efforts are primarily directed toward the alteration of a pathological picture derived from psychiatry and so inevitably lead to our doing whatever we do to every patient in pursuit of symptom relief. Like it or not, psychoanalysis pursues a duality of efforts in both interpreting and allowing psychological development to proceed. We do not have a clear indication for the former save in the somewhat vague concept of "analyzability," nor do we have an exclusive license for the latter.

One way to resolve the proper scientific status of psychoanalysis would be to better clarify the conditions that call for a hermeneutic intervention; i.e., conditions that go beyond analyzability and are distinct from need gratification. However, another case presentation will not prove the point. Analytic interpretive work, in this point of view, is to be considered the sine qua non of therapy, and issues such as need gratification or developmental considerations are seen as preliminary to, or "in order to," pursue interpretation. We listen or we care or we fulfill "in order to" determine what is meant. Sadness, marital conflict or even perverse activity are not to be thought of as "something to be eliminated" as much as something to be understood. This is not meant to denigrate nonanalytic intervention, but rather to position it as lying outside of efforts to understand what things mean.

Objectivity vs. subjectivity, extrospection vs. introspection

It is a given that science must be objective, that is free of expectations, prejudices and personal biases that might influence what is observed and concluded. Yet we have learned, primarily from Popper, that there are no pure observations. They are all impregnated with our theories and our prejudices (Popper, 1994, p. 8). The best we can hope for is to have reached an agreement with a number of equally prejudiced observers. Pragmatism concludes for us that truth is what we can all agree upon and what works. We cannot remove our subjectivity as much as our needing to take it into account.

Observation often is considered to be objective and is cast in visual terms, commonly termed extrospection, which is in contrast to introspection. The latter is the act of gathering information about the inner life of someone and is assumed rather than directly perceived. Thus introspection, or vicarious introspection, is considered to be somewhat less reliable than extrospection, lacking as it does the objectivity of what claims to be visual and therefore apparent. However, vicarious introspection, or the gathering up of information about the inner life of someone else, also runs a greater risk – that of the data that is collected being misinterpreted, inasmuch as it is psychological. It is therefore a hermeneutic discipline. We need to offer a psychological explanation for the feeling of enhanced understanding or accepted meaning leading to a conviction that it is correct and best fits the data that has been accumulated. Introspection needs some form of validation.

The sense of conviction

It is usually easier to reach agreements on "facts," which may be observed, than on states of mind, which are inferred. Vicarious introspection allows for a multitude of meaning, and so consensus can be difficult to attain. Although Kohut (1984) felt that the empathic bond between the self and a selfobject was indicative of a "correct understanding," the necessary added component was a proper interpretation of the meaning of that empathic connection. When the meaning of anything from a dream to a biblical passage is offered to a patient or a reader, the reaction of acceptance is one of compliance if it is accompanied by the qualities of a "good fit," and this most telling quality is that of a sense of conviction. This sense is reflective of a firmness of the self, often explained by a reference to an increased or augmented psychic structure of the self. Therefore, when we move from a misunderstanding of what, say, Kubie is to a better understanding of our misperception of Kubie, we do not see Kubie for what he really is but rather for what he is not. We then gain the meaning of our transference and carry that gain into relationships outside of the analysis. To still need Kubie indicates a weak self-structure and the need for a relationship rather than an interpretation.

The pragmatics of silence

In an informal gathering having nothing to do with psychoanalysis, someone mentioned that he had been in analysis and someone else asked what his analyst had said about something or other. The past patient quickly answered, "well, you know, analysts never say anything" and smiled as he changed the subject. So the "talking cure" of "talk therapy" seemed to rely primarily on the utterances of the patient rather than the verbal outspoken wisdom of the therapist. Psychoanalysis in particular has generated a picture of the silent analyst who may or may not take notes but who rarely reveals what he or she may be noting. Indeed, silence has managed to become a trademark of analysis, one that is sometimes scorned but is both tolerated and accepted as a necessary component of anonymity.

The silent analyst is the unknown analyst, and the rationale for this silence is justified so that the analyst is allowed not to "be" anyone in particular so that he or she can "be" anyone at all. In a sweeping history of silence and Christianity, MacCulloch suggests that we approach and understand the "divine" through negation; i.e., it is necessarily secretive and mysterious (MacCulloch, 2013), and thus not anything of certainty. Not to push the similarity to God to an extreme, it is enshrined in analysis that a revealing of oneself is to be avoided in order to allow the transference(s) to develop. It is not the case that the silence of the analyst merely reflects that he or she has nothing to say. Rather, it is essential that nothing be said so that an uncertain number of attributes can be thought to reside in the analyst. The analyst who says nothing is thereby cast in the role of the analyst who is permitted to think or be everything. This is inevitably made into a convenience of theory that is allowed to make the claim that whatever is attributed to the analyst has its origin in the patient. Thus the analyst is given the position of the interpreter who may safely say, "I am what you choose me to be."

This statement of safety may be an ideal scientific one, inasmuch as it allows for an observation that is in no way influenced by, or contaminated with, a contribution of the observer. The scientist who observes malignant cells under a microscope cannot allow his or her personal inclinations to effect the status of those cells. Observation is held to be pure and without prejudice. The same is hoped for with interpretation; i.e., what the interpreting analyst "is" can in no way or should in no way effect what the patient makes of him or her. Yet we are told by Popper that all of our observations are theory-impregnated (Popper, 1994, p. 58), and so we recognize that only if one knows what malignant cells look like can one reliably observe them and only if one has a workable pre-existing theory can one interpret what is said by another.

Case example

An analytic candidate, Paul, reported that a patient was constantly in a state of anxious agitation, convinced as he was that others were saying negative things about him. He told of visiting a restaurant, talking to the wife of the owner, who asked about him, and soon telling her that he wrote a newspaper column. He saw her in the back room talking to her husband and shortly thereafter proceeding to become busy with a computer. The patient assumed that they were looking him up and gathering information about an unfortunate incident in his past. He was convinced that they now, because of this newly gathered information, would proceed to think negatively of him. Paul reported that he could well understand this agitated state of his patient inasmuch as the man had indeed been some time ago involved in an unfortunate and potentially negative incident.

Paul's supervisor asked him to step back from this position of agreement and to consider why the patient proceeded down this particular chain of explanation. In other words, Paul was asked to interpret the meaning of the patient's agitation, a meaning that went beyond the mere description of events. To do so, Paul would need to employ a theory that might yield a vision of the patient's problem that was not readily available to the candidate, but that might well be offered as an interpretation by a number of persons hearing of this description of Paul's patient. The supervisor asked Paul to concentrate on the need of the patient to feel that he was talked about, but Paul seemed reluctant to move away from the likelihood of the reality of this event; i.e., the patient was correct in his appraisal of the conversation of the restaurant owner and his wife. Once the possibility of alternative explanations was offered and advanced, Paul was able to listen and even to consider how that one explanation seemed plausible, perhaps because he was able to see a similarity in his own psychology. The explanation went beyond the need to feel important, placing the fear of being unimportant onto the need to feel that he occupied a place in another's mind. Paul and his supervisor engaged in the activity of interpretation involving the task of considering a variety of meanings in order to find a "best fit."

From the silent analyst who now speaks in order to offer an interpretation, one moves to the verbal patient who now is called upon to corroborate or refute the interpretation. We recognize that interpretation is bidirectional, or goes both ways, as the patient interprets not only what is spoken by the analyst but also the silence of the analyst. One always interprets what is said as well as what is not said.

The vast literature on silence ranges from Reik's essay on the meaning of silence (Reik, 1968) to Arlow's suggestions on the technique of managing silence (Arlow, 1961) to discussions of the patient's silence (Leigner, 2003) to that of the analyst's silence (Brockbank, 1970). The plethora of writing on silence reflects a variety of attempts to understand and interpret this bidirectionality of silence. If Paul had merely supported and/or agreed with the suspicions of his patient, he would not have enabled his patient to consider what his agitation might mean. A silent Paul would not be functioning as an analyst, but might well be seen as therapeutic. The pragmatic or practical analyst cannot allow silence to define herself or himself inasmuch as each patient employs a parallel theory that is aimed at understanding his or her analyst.

Conclusion

As Kuhn insisted, hermeneutic studies are not the same as empirical studies, although they are still to be considered as scientific. And as Kubie insisted, the interpretations of transference may well leave but little behind. Kubie may have been in error to assume that the analyst would vanish, because one's presence can never be eradicated. The mystery will always remain. Kuhn may also have been mistaken in positing a sharp division between these scientific efforts, because, as we have seen in psychoanalysis, empirical and normative issues are always in consideration. It may be helpful if the critics of psychoanalysis cease the clamoring for controlled empirical studies, and if the critics of hermeneutics cease their claim that they too interpret. It should be no surprise that we are all scientists who both search for truth and settle for interpretations.

Chapter 4

A psychoanalysis independent of psychiatry

Introduction

The title of this chapter might well be "Psychiatry Without Psychoanalysis," for the two fields seem more and more to insist on a separation from a unity that at one time seemed almost indissoluble. Some years ago almost all psychoanalysts were also psychiatrists, and the few psychiatrists who were not analysts were considered to be more or less out of step, rather than representative of a particular position. Not anymore. Indeed, although the requirement long ago for membership in the American Psychoanalytic Association was a medical degree, now it has been modified to a variety of degrees ranging from master's degrees in social work to law degrees and dental licenses. There are many reasons for this continual and continued divide, but there is a lack of clarity as to just what the present relationship between psychoanalysis and psychiatry should be. Is it a permanent separation with no hope of reconciliation? Is it a temporary breach with a modified but congenial connection? Is it but a historical oddity? One approach to an attempt at some workable answer is to consider some different areas of concern and connection.

The trainees

There can be little doubt that the alliance between psychiatry and psychoanalysis was initially predicated on a need for potential analysts to be familiar with all manner of mental illness. Although Sigmund Freud wrote in favor of lay analysis (Freud, 1926a), and there were several outstanding analysts such as Erik Erikson with little or no psychiatric or medical training, in the United States the link between psychiatry and psychoanalysis was able to persist for some time because of the number of available medically trained candidates. Non-medical analytic training was a somewhat unpopular and often discredited path to becoming a psychoanalyst until analytic institutes governed by the American Psychoanalytic Association realized their usual sources of candidates began to diminish and dry up. I personally doubt that it was a realization that one need not be a psychiatrist to become a psychoanalyst that led to a loosening of the admission requirements as much as it was a need to fill the classes.

Of course, psychiatrists are primarily interested in treatment, and Philip Holzman insisted that a concentration on therapy could destroy the more extensive concerns of psychoanalysis. A focus on treatment may well isolate analysis and so refrain from developing the widening of its realm into psychology, jurisprudence and the humanities (Holzman, 1985).

The admission of non-psychiatrists to analytic training does seem to highlight the question of whether there is or need be any connection between these two disciplines. Does the person with a Ph.D. in English Literature suffer from an irremedial training deficit that precludes success as a psychoanalyst, or has the gap between psychiatry and psychoanalysis become so large that the psychiatrist is at a disadvantage when compared to the English scholar, who is more widely read and familiar with the ways of the world than someone who has been sequestered in a hospital for years? Of course, these are but positions of prejudice, but they should direct our thinking to the topsy-turvy relationship between psychiatry and psychoanalysis, one that has moved from interdependence to an almost disdainful independence. Holzman also decries the fact that our institutes are devoted to training therapists rather than scientific investigators who are familiar with psychoanalysis. He comments, "a not inconsiderable number of scientists who have received full psychoanalytic training have entered the full practice of psychoanalysis and thus have abandoned their original intentions and purposes when seeking out training" (Holzman, 1985, p. 766). From the disparate prerequisites for training we move to the disparity in vocabulary.

Language and dependence

There was a time when psychiatry and psychoanalysis spoke the same language. Each seemed to know what neuroses and psychosis meant. Each seemed to agree on the varied diagnoses of conversion hysteria, obsessive compulsive and depression, and even "transference" seemed to be part of a shared lexicon. The vocabulary changed, as did the domain of interest. Just as manic depression became bipolar disorder, the question seemed to be whether this name change reflected an honest-to-goodness different malady or merely a preferred nomenclature or declaration of difference. Whatever the reason, the request of psychiatry was clear; i.e., either you analysts go along with us or you will be left behind as we advance. The situation was not the same, as modifications in psychoanalysis from Lacan to Kohut asked for a similar response from psychiatry. Here the separation became starkly apparent. Psychiatry was changing and could easily jettison psychoanalysis, while analysis had no such freedom. "Progressive neutralization" and "self-objects" had no currency in psychiatry, while psychoanalysis tried vainly to adapt to DSM 5 and its endless categories of illness by plebiscite. No longer were these two disciplines with a mutually supporting network. Psychiatry now ran the show, and psychoanalysis could run alongside or merely drop off. The reasons for this power differential were arguable, but obvious, and ranged from the dominance of health insurance to the popularity of psychopharmacology. The results, however, were not arguable. Psychoanalysis was no longer an equal partner, but rather either a bit of a nuisance or simply inconsequential. The language of psychiatry became the language of mental health and mental disorder.

A colleague involved in the training of psychiatrists told me that a good psychiatrist should be able to "prescribe" both medication and psychotherapy. So too should a good psychotherapist be able to prescribe each or both in a competent manner. If we unpack the word "prescribe," at its best, it should carry the awareness of "the mechanism of action." Thus if a cardiologist prescribes a drug to lower cholesterol he or she has some idea of how it may work. If a valve replacement is indicated, the same requirement of "mechanism of action" is in order. If an antidepressant is prescribed, one should have some knowledge of the role of serotonin and norepinephrine in the origin and maintenance of depression. If psychotherapy is prescribed one should aim for some sort of explanation, ranging from "making the unconscious conscious" or "healing the split" or any of a variety of psychological efforts directed at an amelioration of symptoms. Otherwise the "prescribing" of treatment is something akin to a lottery of chance. The present state of prescribing is very much like that of different languages.

Diagnosis

In May of 2013, the American Psychiatric Association published the fifth edition of the *Diagnostic and Statistical Manual of Mental Disorders*, or DSM-5. This compendium of mental illness had the aim of categorizing and so distinguishing one mental disorder from another. It was and is felt to be imperative in all of medicine for us to make such distinctions in order to eventually arrive at cures for illness. For example, coughing per se is true of many respiratory maladies, but until the *Tubercle bacillus* was isolated and proven to be the cause of tuberculosis, one cause of coughing, there was little that could be done for that illness. In truth, fever charts were kept for years in order to distinguish one respiratory illness from another, but this effort failed until the real causes for the different forms of pneumonia were discovered. Good dermatologists can distinguish one skin disorder from another by mere observation of a lesion, but the same cannot be said to be true for all manner of pain; i.e., kidney stones can be suspected, but the suspicion cannot be confirmed without radiologic evidence. Thus, psychiatry followed some very basic principles in all of medicine in the effort to "know the difference between one condition and another" (Pres, 2013). Of course the reason to know the difference between one condition and another was that of finding the cause and cure for a condition. Coughs due to tuberculosis are distinguished from those due to pneumococcal pneumonia because they have different causal infectious agents and quite different roads to cure. Observation alone will not make the necessary distinction. Thus the effort of delineation and distinction is primarily and indeed solely relied upon to discover the cause and ultimately the cure. The payoff in distinguishing, for example, narcolepsy from other forms of hypersomnolence came from the finding that narcolepsy is associated with hypercretin deficiency and therefore requires a specific treatment (Moran, 2013). It follows that there is a way of distinguishing one condition from another (Hales, 2013). Psychiatry followed in the tradition of all medicine in its investigation of disease and then proceeded to better locate and distinguish these causal factors in the brain. It became neuroscience.

Psychiatry became enamored of neuroscience for perhaps a multitude of reasons, ranging from discontent with psychological causes of mental disorders to effectiveness of pharmacological treatments of such disorders. However, there soon became a reigning conviction in the field that problems of the mind are best thought of as problems in the brain. Psychiatry cannot possibly be faulted for this turn of major concern, and the potential yield in pursuing this effort is the promise of rewards from psychopharmacology as well as a more rapid response in clinical practice.

Treatment

Perhaps because the psychoanalysis of Freud's era was too expensive and/or too demanding of time, a somewhat watered-down or modified form of psychoanalysis inevitably emerged. It was presented as psychoanalytically informed psychotherapy or psychodynamic therapy or even talk therapy. For the most part the field of psychiatry often lumped any and all talk therapies into an amorphous category of psychotherapy and included cognitive behavior therapy (CBT), supportive therapy and psychodynamic therapy into this collection. It often uses the group as a control in the evaluation of one or more psychopharmacology agents. Along with this lack of differentiation of the different forms of psychotherapy, there has been a lack of training in these various, often quite different, forms of psychological intervention, inasmuch as it is a rare person who is competent in all of these disparate forms. Thus psychiatry, in the most general sense, tends to refer patients to one or more preferred providers of psychotherapy. In that manner it takes little or no recognition of psychoanalysis, less of psychoanalytically oriented psychotherapy, and as a result is in no position to aim to pinpoint a particular form of psychotherapy for a particular patient any more than a psychoanalyst might refer a patient to a psychiatrist for a particular medication. Indeed, quite often, psychotherapy in the broadest sense is prescribed only as a court of last resort.

Clinical example

Cases referred for psychotherapy often represent the existing dichotomy of treatment in that the direction of the referral is more frequently from someone who primarily medicates patients to someone who either cannot write prescriptions or else is unfamiliar with the available medications. Thus, a primary care physician or a psychiatrist who is not himself or herself a psychotherapist refers a patient to a psychologist or a social worker often as an act of exhaustion or desperation. Those who cannot medicate likewise refer patients for medication, but probably not as a court of last resort. The statistics are probably not and never will be available, but it would seem that psychotherapists are more willing to seek assistance from psychopharmacology than vice versa. Articles on the efficacy of various antidepressant medications rarely include psychodynamic psychotherapy for comparison and either use CBT or placebos for controls. This would seem to reflect the earlier suggested thesis that today's psychiatry evidences little use for psychoanalytic ideas.

Case report

Kevin is a man in his 40s with a history of alcoholism and depression. He has been sober and abstinent for over a year and is referred to a psychiatric clinic for antidepressant medication. He is seen by a psychiatric resident who tries a variety of medications with no real resultant changes. Kevin is the youngest of three children and was primarily raised by an older brother after the death of his mother when he was eight years old, followed by the death of his father a year later. He said that he was abused by his brother, and he describes a series of events throughout his life punctuated by depression and excessive drinking. The presentation of this patient to a group of psychiatrists is initially directed towards a discussion of the medications and the dosages, although the resident who is seeing Kevin states that the patient loves to come and just talk. The attending psychoanalyst explains to the group about parent loss, and the thesis that, failing adequate mourning, such patients often exhibit a form of emotional arrest after the death of a parent. The resident then exclaims, "why Kevin acts just like an eight year old." The clarity that comes with a psychological comprehension of this or any other patient allows all talk of proper medication and/or adequate dosages to subside. There then follows a discussion of parent loss and the need for the availability of a surviving parent to assist in the mourning reaction, and the role of therapy to re-activate the unacknowledged mourning. Of course, an unfamiliarity with the rather intensive literature on parent loss would restrict one's vision to the chemical imbalance representative of depression. So too would the lack of training in dynamic psychoanalytic psychotherapy disallow an opportunity to manage the grief reaction that an eight-year-old would endure.

Although this case may be illustrative of a form of misapplication of a pure psychiatric approach, there are other cases that demonstrate the folly of a pure psychodynamic approach.

Case illustration

Eloise was seen by an analytic psychiatrist after the birth of her son, which appeared to usher in a severe and unremitting depression. She was very reluctant to consider the idea of medication, and so a course of psychodynamic psychotherapy was suggested. She was enthusiastic about the process and eagerly spoke of her own life, especially of a difficult relationship with her mother. Although the therapeutic sessions seemed fruitful in terms of Eloise's seeming to better understand herself, her depression appeared to be unrelenting. Finally she agreed to begin antidepressant medication, and in a rather short time she experienced relief from her depression, followed by a reduced interest in pursuing psychotherapy. Subsequent pregnancies followed by postpartum depressions brought Eloise back to the therapist, and subsequent antidepressant medication was equally effective. Her therapist wondered if a persistent psychodynamic effort might have been as worthwhile as the medication, and he struggled with the idea of the frequency of the sessions as being a crucial factor. No matter. The psychiatric pharmacologic approach clearly served Eloise best.

Kevin and Eloise represent cases that are not uncommon, yet are initially mistreated for no clear or obvious reasons. Neither case would qualify for the category of gross mismanagement, inasmuch as each was approached with a flexible stance. However, they do seem to highlight different approaches to some mental disorders – those that are handled by psychiatric methods and those that are treated by psychoanalytic methods with no clear differentiation save that of trial and error. As the gap between these disparate approaches to seemingly similar problems became more apparent, and as the cry for a truly scientific investigation became more insistent, there developed an increasing interest in the broad field of neuroscience with the hope that an underlying and indisputable neurological investigation might reveal a commonality between psychiatry and psychoanalysis.

Neuroscience and reconciliation

It seemed beyond argument that there is a biological basis for all of our thinking and acting, so the renewed popularity of the study of the brain seemed an obvious avenue for a reconciliation of the fields of psychiatry and psychoanalysis. There is a certain seductive quality to a grounding of explanation in a biological fact as opposed to the occasional vagueness of a psychological explanation. The argument seems settled if there is clear pathology, say, in the hippocampus or some area of the brain other than in a person's childhood or his/her unconscious – i.e., in the mind.

An interesting exercise in this effort to bridge the gap by way of a more scientific approach to psychological phenomenon is highlighted in a recent book by Baron-Cohen (2011), which is devoted to a careful neurological study of the psychological state of empathy and thereupon aims to explain cruel behavior as equivalent to a lack of empathy. The author devotes a good deal of attention to what he calls "the empathy circuit," which he claims as being made available to science by functional magnetic resonance imaging, and which consists of regions of the brain made up of the medial prefrontal cortex, the orbito-frontal cortex, the frontal operculum, the inferior frontal gyrus, the caudal anterior cingulated cortex and the anterior insular, the temporoparietal junction, the superior temporal sulcus, the somatosensory cortex, the inferior parietal lobule, the inferior parietal sulcus and the amygdala. All in all, there are 10 brain regions involved in empathy, and all seem to play a role in its achievement and practice. Baron-Cohen does note that environmental or other biological factors must also be included in the equation.

This connection between brain areas and a significant form of behavior or maladaptive behavior holds out a promise for joining the vocabularies, the treatment and ultimately the training for those interested in mental disorders. Some difficulties, however, present themselves. They start with language and the definition of empathy offered by Baron-Cohen (2011, pp. 15–16). He would have us focus on what another person is thinking or feeling and respond to those thoughts and feelings with "appropriate" emotion. That word "appropriate" presents a problem. Baron-Cohen has no references to Heinz Kohut and self psychology, but Kohut felt that the use of sirens on the Nazi dive bombers was an example of the Germans' empathy toward the citizens in the city being bombed. They correctly assumed that the shrieking sirens would terrify the citizenry and make them more fearful of the Germans. This was certainly cruel, but could be seen as "appropriate" for the Germans. So too do morticians manage to sell expensive caskets to grieving members of the family of the deceased by being empathic with their grief. The word "appropriate" becomes problematic. In dealing with dying patients, members of "end of life" or hospice teams are often quite empathic while urging patients to choose between prolonging a painful existence or ending it. It is difficult to determine what is the "appropriate" stance in many of these situations. Inasmuch as context becomes such an important and significant issue in so many of these interactions, one is hard-pressed to equate empathy with much more than being a process of data gathering about the inner thoughts and feelings of others. Thus we are challenged about the ease of reducing complex psychological issues with brain activity. The flaw in the effort to reconcile neuroscience with psychoanalysis may be well expressed in the following quote from Paul Ricoeur:

> My initial thesis is that these discourses represent heterogeneous perspectives, which is to say that they cannot be reduced to each other or derived from each other. In the one case it is a question of neurosis and their connections in a system; in the other one speaks of knowledge, action, feeling – acts or states characterized by intentions, motivations, and values. I shall therefore combat the sort of semantic amalgamation that one finds summarized in the oxymoronic formula "The brain thinks."
>
> (Changeux & Ricoeur, 2000, p. 14)

Ricoeur goes on to speak of a duality of perspectives and insists that there is no way of passing from one order of discourse to the other. Thus, if today's psychiatry becomes directed to the discovery of brain disorders to explain disease, then there is no chance of its simultaneously speaking of issues involving thoughts, feelings and motivation.

It is important to realize that the limitations that Ricoeur speaks of are in no way designed to diminish the importance of brain investigations and all of neuroscience, but rather to clarify their domain of interest. The word "appropriate" (that

Baron-Cohen used to delineate empathy) is a word involving values that may or may not involve a particular area of the brain, but essentially it has to do with meaning rather than neuronal connections. One cannot be reduced to the other, and efforts like those of Baron-Cohen are misguided. That they are pursued is testimony to the present-day gap between two fields that were once a union.

Psychiatry and the brain, psychoanalysis and the brain

There can be little doubt that the purpose of psychiatry's present-day effort to be more scientific is to correlate mental disorders with particular areas of brain pathology. This may well be a worthwhile and fruitful pursuit and is in keeping with psychopharmacological advances. The research that is most fruitful in psychiatry does indeed have to do with isolating particular drugs that are most effective in treating the disordered neurochemistry of mental disorders. The future of research into the genetic abnormalities associated with mental disorders is both promising and exciting. Like it or not, it has very little to do with psychoanalysis.

It is difficult to make the case that recent studies in neuroscience have resulted in advances in psychoanalysis. Again, this is not to diminish or denigrate the work of neuroscience, but rather to underscore what Ricoeur calls the semantic dualism of those separate universes of discourse. One universe of discourse is that of the subject in the world of experience. This subjectivity is objectified or treated as an outside object by a variety of scientific studies, one of which is that of the neuronal connections. One good example is that of prosopagnosia, or facial blindness, which is caused by localized lesions in areas of the right hemisphere – i.e., the freeform gynus. Those who suffer from this disorder report a variety of maneuvers that they have devised to aid in recognition, all the time realizing that they simply are unable to tell one face from another. If one compares this rather circumscribed achievement of recognition and discrimination to the feelings that one has at meeting a long-lost relative or lover, one is able to see the difference between the mere recognition of a face and the meaning that a particular face has for you. This difference between an objective registration and the complex issue of meaning is the difference between a field that searches for clear facts and one that investigates how we react in a manner that may not easily be circumscribed or demarcated – i.e., what things mean to us.

To paraphrase one philosopher:

> Meaning is wider in scope as well as more precious in value than is truth . . . But even as respects truth, meaning is the wider category; truths are but one class of meanings, namely those in which a claim to verifiability by their consequences is an intrinsic part of their meaning. Beyond the island of meanings which in their own nature are true or false lies the ocean of meaning to which truth or falsity are irrelevant.
>
> (Dewey, 1939)

Transference configurations are not mere misperceptions of others but are, when experienced in psychoanalysis, complex relivings of what another person means (or meant) to you. A recent issue of *Behavioral and Brain Sciences* (Lindquist et al., 2012) states that the locationist approaches to emotional categories are in error, and that emotions are maintained in more general brain networks. That might be read as holding out a promise that one day neuroscience will be able to better correlate the complexity of transference to such networks, or it may be seen as validation of Ricoeur's premise of two separate universes of discourse. He warns us of reducing the one to the other and thus losing the essence of meaning.

The twofold approach to emotion consists of the locationist account and the psychological constructionist account. The first assumes that there are emotion categories such as anger or sadness that can be localized to discrete brain locales. The second assumes that the categories of emotion emerge from the combinations of psychological operations that are not specific to emotions. Rather large-scale networks interact to produce psychologic events. Lindquist describes the various parts of the brain that are said to be the basis of emotions, such as anger, sadness, etc., and shows how most, if not all, of these connections are in error, and that multiple brain areas participate in a psychological construction that gives meaning to a complex state (Lindquist et al., 2012, pp. 121–142).

Silvan Tomkins said in 1981:

> What we ordinarily think of as motivation is not a readily identifiable internal organization resident in any single mechanism but is rather a very crude, loose approximate conceptual net we throw over a human being as she or he lives in her or his habitat. It is as elusive a phenomenon as defining the locus of political power in a democracy. . . . [I]t is everywhere and nowhere and never the same in one place for very long.
>
> (Tomkins, 1981, p. 321)

Sohms and Turnbull (2011) give an example of what they term the dialectical approach in their observations of patients with right parietal lesions who also exhibit self-deception. These patients are paralyzed on the left side, but insist that they are not paralyzed and explain away their paralysis through "transparent rationalizations." Essentially these patients deny a certain segment of reality. The authors proceed to explain how a psychoanalytic approach allowed them to "observe the dynamic phenomena revolving around emotional states." They then employ one or more psychoanalytic theoretical devices, such as "narcissistic defensive organization," to further explain and contribute to what they call "behavioral neurology."

Of course, denial and disavowal have long been known and studied in psychoanalysis, and one wonders if much is added by a neurological perception. We must also couple this question with the above-mentioned fact that the locationist approach is in error, and so right parietal lesions or any such effort to localize the substrate of the phenomena of denial may be foolhardy. I fail to see how Sohms and Turnbull offer anything more than an anatomical fact of interest. In fact, great strides have been made in the psychoanalytic study of disavowal without the benefit of knowledge about the parietal lobe (Goldberg, 1999). One may well wonder if Sohms and Turnbull are themselves limited in their comprehension of disavowal in a psychoanalytic sense and so unable to appreciate the "splitting" that such patients employ. The dialectical approach needs to employ a dual approach.

It may be worthwhile to comment on the attitudes that often impede the hoped-for conversations that could take place between psychiatry and psychoanalysis, along with the many forms of the latter. Although only a few analysts and psychotherapists are authorized to prescribe medication or are familiar enough with this new and expanding field, many psychiatrists feel equipped to do psychotherapy. It is not uncommon to see or hear of patients who claim to be in psychotherapy that consists of little more than friendly listening or advice on how to live. Psychiatrists these days often have completed a few didactic courses on psychotherapy along with CBT, but with have done little work on problems of countertransference or much reading of the psychoanalytic literature. Analysis is rarely advised ostensibly because of the time and money involved, but perhaps more so because of the ignorance of the field. Efforts to allow non-medical analysts or therapists to prescribe medication are actively and vigorously fought by physicians in the various legislatures. Efforts to train psychiatrists in the practice of psychotherapy are not the subject of controversy, although the danger to a patient might well be of equal impact. "Let the consumer beware" should be true of both cases.

An exercise of difference

It is fairly well accepted now by those in mental health that depression can be effectively treated with antidepressant medication or with psychotherapy with somewhat equal effectiveness, but with a higher rate of effectiveness by both. The STAR*D research study of a large number of patients with depression demonstrated that one cannot predict which drug will work or how long a particular drug should be administered to determine effectiveness (Gaynes et al., 2009). The above-mentioned cases of Kevin and Eloise were examples of cases that could either be helped with medication or with therapy, but needed to be matched correctly. If depression is a unitary and definitive illness, it is difficult to explain how it is sometimes helped by one form of treatment and sometimes by another and better by a combination of the two. If the brain is a substrate of the mind one should be able to correct the pathological neuronal connections with a treatment that does just that, or if the mind is a complex subjective disordered state then one should be able to correct the disorder with a talk therapy that does just that. How then to explain that sometimes one works and sometimes the other, and sometimes both do the best job?

A recent research project conducted at a local hospital consisted of the presentation of a patient (chosen more or less at random from an outpatient referral) to a group of therapists who specialized in psychodynamic psychotherapy, cognitive behavioral therapy, psychopharmacology or group therapy. No doubt other modalities of treatment could have been included, but the hope was for a discussion that might reveal a process of thinking that would clarify how best to decide what would serve a particular person best. What was exposed was that the decisions reached had to do with agreement with the person with the most successful power of persuasion rather than the emergence of what treatment was indicated for a particular patient. For the most part, people do what they do and persist in it as long as they can. Some therapists do talk with other therapists who practice differently but, more often than not, these conversations take place after failure rather than as part of a diagnostic assessment. Undoubtedly these conversations routinely involve different languages spoken by therapists with different backgrounds and prejudices. One step toward a solution would be that of promoting individuals who were themselves familiar with all the modalities of treatment, but that is now more hope than reality. As long as this tribal warfare remains, there can be little likelihood of the development of a third discourse – one that unites psychoanalysis and psychiatry, one that can determine which patient with clinical depression will benefit from medication, which from psychoanalytic psychotherapy, which from neither and which from both.

Discussion

Seeing psychoanalysis and psychiatry as partners in the study and treatment of mental illness, joined by a neuroscientific investigation of the workings of the brain, is a not uncommon stance of many present-day therapists and researchers. Seeing psychoanalysis and psychiatry as quite distinct fields of study, research and treatment is a challenge to that partnership. The resolution of this difference will not be brought about by either debate or discussion but will most likely be delivered by the forces of evolution. We shall all be carried along by whatever succeeds, and we shall enjoy the pleasure of choosing right or the dismay of choosing wrong without really having much to say about the eventual mutation that we might well not even recognize. Although we may participate in the evolutionary process and flow, we really cannot do much more than recognize and perhaps wonder at the often unexpected results. Thus far it seems fair to say that psychoanalysis and psychiatry are growing apart, and we would do best to hope for an intact right parietal lobe that does not attempt to deny the obvious.

In his latest book, Edward D. Wilson (2012) writes, "human beings are actors in a story. We are the growing part of an unfinished epic." He also makes the claim that group selection involving cooperativeness, empathy and patterns of networking is the driving force of evolution. Wilson proposes a theory called eurociality in which the biological organization has group members staying together, cooperating and dividing labor. He decries movements that are divisive and non-altruistic and selfish. However, we are now in a period of opposing groups – or perhaps a better word would be tribes – that downgrade other-group members and take pleasure in the downfall of enemies. That may explain the growing enmity between psychiatry and psychoanalysis. A new group will ultimately evolve out of the multiple competing hypotheses, but until then Wilson tells us that group versus group warfare is ingrained in us.

Implications

There are both positive and negative aspects to the union of psychoanalysis and psychiatry and, not surprisingly, the same applies to the dissolution of that union. Having psychiatry as a partner has lent psychoanalysis both prestige and protection. The status and often associated fantasies of idealization attached to medicine did serve for an easy acceptance of analysis, but the problem of lay analysis was an early potential disruption of that alliance. However, the association with psychiatry also meant an inevitable association with mental illness and treatment, and one must wonder if that has been an advantage for psychoanalysis. Despite Freud's initial presentation of analysis as a treatment of mental illness, there can be no doubt that a conviction that "everyone could benefit from analysis," along with the stipulation that all potential analysts should be analyzed, weakened the link between psychoanalysis and psychopathology. Either we were all sick or we could all use an analysis, no matter the presence of mental illness. Once freed of the status of stepchild to psychiatry, perhaps psychoanalysis might consider the options of either complete independence or an allegiance to another discipline.

For starters, imagine a psychoanalysis free of the implicit requirement to follow the psychiatric preoccupation with disease categorization. Analysis might choose to organize its own manual of disease based upon psychoanalytic theoretical constructs, or it might work toward an eventual form of self-understanding that did not have a disease model. Perhaps there are psychological states that call for insight and understanding that have no real relevance to illness, and perhaps there are psychological disorders that do not fall at all into the psychiatric arena of concern.

It may be difficult to imagine a psychoanalytic form of inquiry that did not direct its attention to symptom, inasmuch as a discipline that is solely devoted to illness yet claims itself to be a member of the family of psychoanalysis has blossomed over the years. This supposed blood relative of psychoanalysis is called psychodynamic psychotherapy and may or may not be practiced by psychoanalysts, but is surely one of the predominant forms of psychological intervention employed in the treatment of mental illness. Here is where the single-minded devotion to the categories of mental illness formed by psychiatry is brought to the attention and concern of a practice that may be a far cry from the psychoanalysis introduced by Freud and taught in our psychoanalytic institutes, but one that has gained popularity by avoiding the pitfalls and negative associations that classical psychoanalysis has in terms of cost and time. The question of whether psychodynamic psychotherapy has brought benefits to psychoanalysis or has allowed classical analysis to be dismissed or downplayed presents itself as a result of this linkage between psychiatry and psychoanalysis. Might analysis have profited from an independent existence outside of psychiatry or as a member of another scientific activity had it insisted on a set of internally consistent principles of training and technique that are not a part of psychodynamic psychotherapy? Can a nonanalyzed psychiatrist practice an analytically oriented psychotherapy, or is that asking too much of such a therapist? Of course the answer to that question exists in the many forms of therapy that are often called psychotherapy but are quite alien to both psychoanalysis and psychodynamic psychotherapy and that ask no personal treatment of its practitioners.

Following the divorce of psychiatry from psychoanalysis, the new list of potential suitors and substitutes did not only claim freedom from psychoanalysis but some members of the list were often disdainful of a preoccupation with the problems of the practitioner. The list is long and includes cognitive behavioral therapy, dialectical behavioral therapy, interpersonal therapy, mindfulness and varieties of combinations of medication and psychological interventions. Although the replacements for psychodynamic psychotherapy are essentially psychological and usually based upon some form of "talk therapy," they are open to criticism as lacking a theory as to their effectiveness; i.e., they are offered on a purely pragmatic basis. It is sufficient that they are effective. Of course, that efficacy is rightfully measured against the signs and symptoms of mental illness and not against the criteria of a psychoanalysis based more upon self-understanding. It seems to be true that psychoanalysis may play a small part in the treatment of mental illness, but that it is not in its best interests to be totally defined by that role.

Summary

Psychoanalysis and psychiatry have evolved over time into separate areas of action dealing with mental illness. They have little in common, as noted in their training, their language and their manner of treatment. Efforts to unite them based upon the underlying biology of the brain have not been successful, and over time the two fields may well become more adversarial and distant. These powerful forces of group selection and group cohesion make for a form of warfare that is being waged today in the arena of choice of treatment. Unification may be unlikely because the opposing groups are engaged in difference discourses. The ultimate solution to a workable discourse about mental illness may well be presently beyond our imaginations, as is regularly the case with evolution. Some of us may live to see the new discourse. For now we must recognize a more profound difference between these disciplines based upon their ultimate aims. It is not a question of whether psychoanalysis is a branch of medicine, as is psychiatry or a branch of psychology (Brandchaft & Stolorow, 1985), but rather a need to clarify the very distinctly different goals that these two efforts seek to fulfill. As a start we can turn to the very concept of illness and to the manner in which psychoanalysis approaches mental disorders.

Chapter 5

The concepts of normality and psychopathology in psychoanalysis

Introduction

Normativity is the word applied to the study relating to norms, how they are realized or discovered, how they change or resist change and especially how they impact individuals and society. Psychoanalysis might well be expected to have a significant voice in this discipline, as it speaks to how someone is categorized as being psychologically normal or abnormal. For example, Sigmund Freud as founder of psychoanalysis had no difficulty in one declaration of what is normal when he championed heterosexuality as the sine qua non of normal behavior, both sexual and otherwise (Freud, 1905b). His daughter, Anna Freud, offered another paradigm of normality in her presentation of development being a normal pathway to ultimate maturity, i.e., normal behavior equals mature behavior (Freud, 1965). Ernest Jones lists four criteria for the supposed state of normal. They are drive taming, transcendence of narcissism, capacity for enjoyment and self-contentment, along with freedom from anxiety. Jones did not believe all of these characteristics would be found in one person, but held these out as ideals. We can see that this definition is a mixture of analytic ideas and more descriptive general ones. Melanie Klein offered five criteria for assessing "optimal integration." They are (1) emotional maturity, (2) strength of character indicated by lack of splitting of objects, (3) the capacity to deal with conflicting emotions, (4) a balance between internal life and adaptation to reality and (5) a successful welding into a whole of different parts of the personality (Offer & Sabshin, 1984).

One can readily see how these definitions of normality come from both psychoanalytic data and the particular theoretical orientations of the authors, plus a good deal of what might be termed a folk psychology perspective, as seen in the phrase "emotional maturity." However, the *Textbook of Psychoanalysis* states that psychoanalysis has no definitive position on normality, and so we are presented with a dilemma (Gabbard, Litowitz, & Williams, 2012, p. 578).

It may be of some interest at the onset of our discussion to recognize that some scientists feel that "ideally" science should be descriptive rather than normative (Elqayam & Evans, 2011). Norms are seen as "oughts;" i.e., they are evaluative, whereas descriptions merely say what is. Norms follow rules and laws whereas descriptions avoid bias and rely only on evidence. Of course, once we consider pathology we depart from what "is" to what "ought to be." However, pathology need not become a slave to morals or opinions. One example of the lack of clarity about delineating pathology is the phenomenon of gambling. We do differentiate the pleasure that some people get from gambling from the psychological malady termed pathological gambling. Here is one point where we may turn to psychiatry and its consideration of pathology. For psychiatry there are rules and laws for normativity.

Norms and psychiatry

Our psychiatric brothers have long struggled with the problem of describing and delineating the abnormal of psychology or the components and categories of psychopathology. The solution to the problem of just what is to be considered abnormal has given birth to the publishing and refining of the DSM-5, or the manual of mental disorders (DSM-IV) that codes psychopathology according to the International Classifications of Diseases (ICD). This coding makes an effort to categorize mental disorders in order to achieve what is called reliability, or a set of agreements that aim to communicate about, study and treat people with various mental disorders.

Lacking a fixed and certain foundation for claims of abnormality in neuroscience – i.e., there are no neatly demarcated areas of pathology in the brain for mental disorders – psychiatry often adopts a moral tone in its categories of disease. The fifth edition of the DSM lists six personality disorders, and phrases ranging from "lack of concern for others" to "reluctance to pursue intimacy" to "mistrust and neediness" to "attention-seeking" to "unreasonable" to "unusual beliefs and experiences" are employed in the descriptions of these disorders. The underlying assumption in this form of categorization is one that involves a norm of desirable traits that becomes associated with health and raises the "oughts" of normality to extremely vague characterizations such as "likeability" and "niceness." Essentially we dictate how people ought to behave. Pathology becomes a form of social sanctions.

Psychoanalysis seems to have a much easier time of categorizing pathology inasmuch as it, in its classical form, considers unconscious activity to be responsible for all symptoms and disorders. Freud may be seen as more interested in treatment than in diagnosis, and his cases are not neatly classified into familiar DSM categories. Making the unconscious conscious is the aim of analytic treatment, and the post-Freudian changes, especially those of the ego psychologists, allowed for adaptation to be a significant factor in judging abnormality (Hartmann, 1939). It seems fair to conclude that the many alterations and modifications in psychoanalysis from Klein to Kohut to Lacan to relationist, etc. all focus on the nature of the analytic relationship to adjudicate normality and pathology, and so the transference configurations become the arena for determining normal. While the subjectivity of general psychiatry is controlled by the rules of classification of the DSM, that of psychoanalysis is regulated by the continual analysis of the countertransference, which need not be considered itself pathological.

Marcia Angell (2001), in a cogent and striking review of recent books dealing with this topic, cautions us to note that reliability (the basis of the DSM) is best thought of as consistency or agreement, whereas validity refers to correctness. The classifications offered by psychiatry are directed toward only a consensus of sorts rather than what others might consider a scientific foundation (Jacobs, 2011). Angell quotes Robert Whitaker, who dates the push for psychiatry to become more scientific in 1977 when the medical director of the American Psychiatric Association (APA), Melvin Sabshin, declared, "a vigorous effort to remedicalize psychiatry should be strongly supported." As a result, the APA launched an all-out media and public relations campaign to do just that (Angell, 2011, p. 20). Sabshin was himself a psychoanalyst and saw no conflict or even much division between what has recently become a significant cleavage between psychoanalysis and general psychiatry. Not until the pharmaceutical industry with its emphasis on drug treatment became dominant did the divide between any form of a talking treatment and the growing variety of drug treatment become so blatant. Indeed, the oft claimed support for this divide is the now popular interest in neuroscience, which holds hope for the field of psychiatry to be welcomed back as a bona fide medical specialty based upon seeing scientific principles – i.e., demonstrable changes in the brain. Such a foundation would move away from mere consensus to something approaching validity.

David Jacobs (2011) joins with Angell in decrying the pursuit of psychiatry in its classification efforts to be truly scientific as an exercise in folly. Angell quotes Daniel Arat in discussing this foolishness:

Patients often view psychiatrists as wizards of neurotransmitters, who can choose just the right medication for whatever chemical imbalance is at play. This exaggerated conception of our capabilities has been encouraged by drug companies, by psychiatrists ourselves, and by our patients' understandable hope for cures.

(Angell, 2011, p. 21)

Jacobs goes further than Angell in declaring that the diagnosis of a mental disorder is essentially a judgment much like "this is pornography rather than literature," and so is based on an opinion of a clinician rather than what might be considered "scientific." He says that a technical scientific term should have a specific, objective, value-neutral, non-contextual meaning such as "pneumonia," while the DSM makes the claim that mental disorder lacks a consistent definition that covers all situations and so eludes these conditions.

Jacobs underlines a specific issue in his attempt to differentiate technical scientific diagnoses from opinions. That issue is the distinction between an agent and a host. The first requires recognition that a person is portrayed as someone capable of describing what he or she is and how he or she behaves. The second – i.e., a host – is someone who is harboring something foreign or alien. In the book *The Analysis of Failure* (Goldberg, 2011a), the author refers to a paper defining "treatment resistant" depressed patients as those whose depression seems to resist SSRIs with no consideration given to the individual psychology of these patients. The particular author (Brent et al., 2009) of these articles seems to see depression as akin to tuberculosis, a disease that is susceptible to certain antibiotics and resistant to others. Jacobs alerts us to the fact that if the person as agent is ignored then travesties like the antebellum diagnosis of "drapetomania" (the mental disorder that caused slaves to flee to the North) can and will result (Jacobs, 2011, p. 67).

This distinction between agent and host is not to deny that all of the mind is correlated with the brain. Rather it is to emphasize that considerations such as norms are to be thought of as social practices; i.e., as "oughts" of behavior and not as disorders of neurotransmitters. Alzheimer's disease is certainly a brain disorder, but slaves escaped to the North because of the misery they endured.

Norms and psychoanalysis

We have noted that norms are thought of as "oughts" in terms of what ought to be, how one should think and how one should behave. Judith Jarvis Thomson (2008) divides oughts in two; i.e., those that are directive, as in the statements "A ought to be kind to her brother" or "B ought to get a haircut," and those that are evaluative, as in "that is a good toaster." In psychoanalysis we often see this division in terms of moral and sexual categories. When a book on morality in psychoanalysis (Goldberg, 2007) was reviewed (and I shall not reference the reviewer), the reviewer was upset that the author admitted to not wishing to spend any time talking to an ex-patient that he happened to run into at a large gathering. The reviewer felt that the author was not a good therapist; i.e., he had a moral failure in spite of the fact that the entire book was devoted to challenging a variety of moral positions regularly attributed to therapists and analysts. In the sexual sphere, a book entitled *Love Relations: Normality and Pathology* (Kernberg, 1995) seems to be a manual of what is felt to be normal and abnormal in sexual activity. This capacity to determine normal is both directive (i.e., one should be heterosexual) and evaluative (i.e., it is sick to be promiscuous) and is based upon this author's theoretical tools of diagnosis and treatment. Indeed, many efforts at delineating abnormality depend upon utilizing one's own particular theoretical outlook.

If we go beyond individual theories to a more general perspective on how psychoanalysis considers psychopathology, we can find a range of opinions from those that seem to suggest that psychopathology is rather obvious (McWilliams, 1994) to those that claim that psychoanalysis has nothing to do with psychopathology. Sadly, the aforementioned *Textbook of Psychoanalysis* has no index listing whatsoever for diagnosis. One author seems to equate psychopathology with suffering (McWilliams, 1994, p. 143) while Gitelson (personal communication) felt that psychoanalysis was an effort to bring about self-understanding with any form of symptom relief which is best seen as a valuable but by no means essential byproduct of the endeavor. Some of the problem of not being more definitive lies in the conviction that surely everyone can utilize psychoanalysis (i.e., may benefit from it), but not everyone will necessarily need it (Caligor et al., 2004). It is perhaps comparable to the advice about losing weight – i.e., it is frequently a good thing to do but falls short of one being labeled as obese, or having an illness.

Case example

Marvin came to see a psychoanalyst because a group of his associates at an office that he was in charge of held a meeting about their dissatisfaction with his leadership and proceeded to urge him to see a psychoanalyst. They told Marvin that they were worried that he might be depressed, but Marvin felt that they were primarily displeased with his leadership and that this particular group feeling stemmed from the group's worry about money. He was more than happy to talk to an analyst but he denied any sort of difficulty. He was happily married, the father of two and had no history of psychiatric illness. Inasmuch as the analyst felt that some patients come for treatment without a clear chief complaint or indication for therapy, he decided to conduct a prolonged diagnostic evaluation. Alas, nothing seemed to emerge. One could see how some of Marvin's associates might dislike him, but "likeability" hardly seemed to qualify one for treatment. Marvin did not seem to fulfill the rationale regularly offered for the analysis of analytic candidates whose vocation seemed to require a personal analysis. The problem could well be summed up in a question that asked when and if psychoanalysis should be considered, both as to whether it would be effective as well as to exactly what was to be the nature of the pathology that it was to treat. That seems to be the core question that allows one to say that analysis has no position on normality. That is, if everyone can benefit from it, does that mean that everyone is ill?

This latter part of the question presented itself to me in a different form with two patients. The first was confined to a phone call from a mother of a teenage boy who said that her son wished to be a girl. She had heard that I had an interest in sexual perversions and wondered if this area of treatment would include her son. I quickly answered that I would be happy to help him to better resign himself to being a man, but that I could not help with a sexual transformation. She asked her son and called back to say that he was simply not interested in what I had to offer. I clearly remember my own conviction that this possible patient was sick, that he suffered from some sort of a mental illness and that any treatment of whatever form should be directed toward what I had offered. Years later I changed my mind. In her introduction to an issue of *Psychoanalytic Dialogues* devoted to the subject of "Transgender Subjectivities," Virginia Goldner writes, "we are still drawn to taking normative goals as the superior outcome, even though it hurts and is harmful" (2011, p. 154). I could see just how much it must have hurt that boy to feel that people such as myself did not like what he felt he was. He surely suffered, not necessarily from a mental illness, but rather from those others who felt him to be abnormal.

The second patient that brought me back to the question of where and what psychoanalysis struggles with in its role and its effectiveness was a university undergraduate who came to see me some time ago after a love affair gone bad. This patient was a homosexual young man, and I rather automatically considered him to be sexually abnormal as well as depressed from his thwarted love affair. Fortunately or unfortunately my patient shared none of my prejudices about his sexual orientation, and while I continued to construct an assortment of psychoanalytic theoretical explanations for his supposed pathology, my patient got better from merely talking about his life, including his love life. Inasmuch as I could make no inroads into the arena of abnormality that I had constructed, I went to an analytic supervisor for help in my effort to better understand and so cure my patient's homosexuality. Alas, supervision was of little help beyond urging me to persevere, and ultimately my no longer depressed gay patient terminated happily. In retrospect there was no analytic evidence for this supposed abnormality.

Over the years homosexuality has become depathologized, and both psychiatry and psychoanalysis have had to reckon with their error of judgment. Some years ago in an article about this process of depathologizing, the author quoted Robert Michels, who had insisted that psychoanalysis simply has no position of its own regarding psychopathology (Goldberg, 2001a). This fact leads naturally to the ultimate question of how we could make such an error about sexual preference. My earlier conclusion was that homosexuality is not, properly speaking, a psychoanalytic datum. However, we must remember that using development as the standard of normality allowed Anna Freud to see homosexuality as a developmental stage on the road to mature heterosexuality. Can it be that neither homosexuality nor heterosexuality can be a province of pathology any more than can any trait, and thus only in an individual person can we determine its meaning (Goldberg, 2001a, p. 1113)? Now I believe we can go further to see both the implications of our shortsightedness and the potential psychoanalytic contribution we can make to the study of norms.

Implications

A thought experiment should allow one to better comprehend how those two patients must have felt about being seen as different or abnormal by others. Just as a child experiences the recognition and acknowledgment of the mother or "the gleam in the mother's eyes" (Kohut, 1971), the child likewise sees the disappointment or outright rejection of his or her self by the failed mirroring of the parent. Isay (1989) has noted this in the father's rejection of the advances of the homosexual boy, but surely all sexual presentations that fall outside of a societal ranking of norms will meet a similarly negative fate. A parent may not easily welcome certain traits in a developing child if the hope is for an athlete or a musical prodigy or even a blond, blue-eyed baby. However, traits are usually, but not always, felt to be God-given or inborn or unmodifiable rather than willful or deliberate. The radical evaluation of homosexuals or transgender persons or even transvestites may be to see them as possessing certain traits such as having red hair or being tall rather than as being sick or deviant. Resistance to the usual moral classifications of abnormality might allow psychoanalysis to stake out its own standards of normality and pathology.

Psychoanalytic contributions

Freed from the norms inherited from either classical psychoanalytic culture or from collected psychiatric prejudice, today's psychoanalysis should strive to develop a pathology that is derived from the data of psychoanalysis. These data are the transference configurations and not traits such as red hair or gender or even some traits such as shyness (Lane, 2007). Indeed kindness may be an admirable trait, but it need not be a sign of health any more than a cold and aloof individual is to be considered sick. Certainly some analyses result in personality and character changes that are considered desirable, but we would be well advised to be cautious in considering all such change as indicators of health. Just as boys of short stature may be given growth hormone with the goal of making them taller, we do not consider shortness as a sign of pathology.

Some clinical cases do lend themselves to a clear categorization of sickness and health based upon certain clear transference material, and among these are a group of behavior disorders. Inasmuch as "behavior" is readily drawn into a societal approach of pathology as seen in misbehavior such as stealing or sexual promiscuity, an analytic perspective may aid in the determination of pathology that goes beyond overt classification.

Case example

Cases of shopping sprees are good examples of the need to delineate the normal from the abnormal. Sally was a financial consultant who had risen to a prominent position in a consulting firm despite her often fraudulent itemization of her expense account and her deep debt due to sporadic and erratic shopping sprees. Sally had a close friend who also enjoyed shopping and who occasionally shopped to excess but who never went into debt nor shared the characteristics of Sally's shopping. Sally's sprees were conducted in an almost dissociated state and regularly culminated in the purchase of items of clothing never used or even contemplated for wearing. Sally had been to many psychiatrists and group therapy sessions to no avail until she entered psychoanalysis.

The particular transference configuration that characterized Sally's treatment has been described elsewhere (Goldberg, 1999) and is termed a vertical split. The self-organization of Sally is best seen as divided with one sector, the reality ego, set apart from another that is grandiose and impervious to the constraints of reality. Sally would return from a shopping spree with a hearty dislike of herself for her uncontrolled behavior and would either hide her excess shopping from her family or beg their forgiveness. When these behavioral segments occurred in the analysis, the analyst was similarly torn in an attempt to be empathic with both the contrite Sally and the disavowed shopper. The countertransference reactions of the analyst were of a corresponding nature in that one might wish to control or discipline Sally for her excess while one might also be able to identify with the excitement and pleasure of shopping.

Not surprisingly Sally's split began in childhood and seemed to echo the split in the acrimonious divorce of her parents. Sally struggled with the extremely enjoyable visits with her father, which had to be hidden from her mother primarily in terms of the pleasures she experienced when her father took her shopping. One can organize one's thinking about Sally's psychological makeup in terms of conflicting identifications or all manner of theoretical rendering, but there is a clear and cogent psychopathology that represents Sally in contrast to her non-conflicted shopping friend. Sally's sprees are erratic and are retrospectively disliked and often felt as alien. In her analysis the onset of the sprees was regularly initiated by a promotion or even a compliment on her performance, which seemed to stimulate an unregulated, grandiose fantasy of pride and exhibitionism, which only later became available for scrutiny and regulation in the analysis. The success of Sally's treatment was predicated on healing the split, joining the separate sectors of reality and resulting in an integrated self. Here we may begin to see how the transference configuration could be seen as abnormal or deviant and so represent pathology, just as Sally's behavior could do the same. Sally became the agent in charge of her behavior when she healed her split.

Case example

Behavior disorders present a variety of transference and corresponding countertransference configurations that need not be readily conceptualized upon initial presentation. This serves to underscore the fact that a purely phenomenological or descriptive evaluation of a patient need not reveal the underlying pathology. It also holds that one cannot easily recognize the disavowal that so uniformly is the sine qua non of the vertical split.

Eric was a physician referred to an analyst by a treating psychiatrist who feared Eric's infidelity to his wife, coupled with sporadic sexual encounters with a variety of patients, would get Eric into significant legal difficulty. Eric felt that this psychiatrist was very supportive and sensitive but could only warn him of the dangers with which he flirted lest a patient file an ethical complaint against him. Eric himself felt that his dalliances were perhaps "stupid," but he also felt that he was doing what any man locked in an unhappy marriage would do; i.e., he felt no guilt or remorse, qualities looked for in Eric by the referring psychiatrist. Essentially Eric did not feel bad in any manner that corresponded to what this initial therapist looked for.

Without going into the details of the analytic treatment of Eric, suffice it to say that under the bravado of Eric's rationalizations of this errant behavior a profound dislike of these furtive escapades did emerge.

Eric's analytic treatment was characterized by an initial enthusiasm manifested by a wish to come more often that was voiced but not realized. That was followed after some more analytic work by a wish to come less often. Together we likened Eric's struggle with choice and decision to his recent purchase of a new car, which seemed to exemplify his inability to gain satisfaction. As much as he wanted a particular style and model of one automobile, he ended up purchasing one that was merely adequate and not what he really wanted. He said that that was like his marriage, but also quite representative of his mother. She was a paradigm of disappointment and dissatisfaction. Eric said that the atmosphere in the world of his childhood was the conviction that others were able to obtain all manner of things, while his own family always fell short save for brief periods of short-lived pleasure. These spurts of joy were both rare and secretive and felt to be undeserved. Analytic work enabled Eric to see his sexual acting out with patients as something that he indulged in with guilt and remorse yet also something he felt that he deserved. His so-called misbehavior was a complex collection of disavowed actions that qualifies as a pathology that would be so classified only on the basis of the transference and countertransference evidence.

One must be wary of the assumption of any behavioral manifestation becoming a bona fide indicator of psychopathology. Sporadic sexual behavior of either homosexual or heterosexual form may be signifiers of psychopathology if the transference displays the aforementioned split and disavowal. This demands a history of the behavior being repetitive and felt as alien to the remainder of the psyche. In summary, it is not the behavior but its place in one's psychic organization that bespeaks abnormality.

Abnormal transferences

Aside from the more or less striking evidence of a vertical split in behavior disorders, psychoanalysis can correlate psychopathology and the transference manifestations that emerge in analysis with other forms of pathology. The various theoretical approaches and "schools" of psychoanalysis no doubt have their own considerations as to what constitutes abnormality in the assessment of transference. A recent issue of *Psychoanalytic Dialogues*, which is presented as a journal of relational perspectives (Goldner, 2011), is devoted to the topic of what is called "Transgender Subjectivities" and clearly struggles with the problem of abnormality in gender. Ken Corbett, writing of Gender Regulation, notes that "too often analysts have looked at variance and called it illness" (Corbett, 2011, p. 457), thus highlighting the fact that gender is not a psychoanalytic datum, and a relational perspective is striving to yield a clear differentiation of variance from illness.

When Heinz Kohut (1971) introduced some new forms of transference to the psychoanalytic lexicon he offered new categories of psychopathology based on these transferences. He insisted that narcissism was not in itself to be equated with pathology no matter how pejorative a connotation it often carried. Rather he delineated specific narcissistic disorders that were characterized by unique forms of transference, termed selfobject transferences. Of course his theory explicated these forms and so allowed for this categorization of illness. These transferences emerged in analytic work and were often able to correlate with some overt behavior, but true pathological demarcation could classify narcissistic disorders along a spectrum from primitive to mature in accord with the particular transference manifestations rather than that of the overt behavioral issues, no matter how accurate the phenomenological appraisal might be. These descriptions are not the material of psychoanalysis. They are often merely the "opinions" noted earlier.

Another example of using emerging transference and countertransference to be representative of psychopathology can be seen in the object relations explanation of borderline personality disorders (Kernberg, 1976). Although there is today no clear consensus on the psychological organization of this syndrome there does seem to be a general agreement on the countertransference issues that arise in the treatment of the disorder (McWilliams, 1994, p. 90). These are often primarily descriptive but can be easily translated into the terminology of object relations theory as well. Indeed many patients are diagnosed as having borderline personality disorders because of their impact on the diagnostician; i.e., the countertransference regulates the treatment.

The advantage of psychoanalytic diagnoses

There can be no doubt that many patients fit well with psychiatric diagnosis and so do well with the appropriate treatment. A psychoanalytic diagnosis may be time consuming, costly and of no apparent benefit. However, this is not always the case.

Case example

Gerald was an elderly man who became depressed following the death of his wife after 40 years of what was described as a simply wonderful marriage. A referral to a psychiatrist quickly established the diagnosis of clinical depression due to a severe mourning reaction, and a course of antidepressants was prescribed. The first antidepressants used seemed not to be effective and in keeping with the usual procedure of trial and error in the employment of antidepressant medications a series of such pharmacological interventions was initiated, all to no avail. This was followed by a variety of other non-pharmaceutical treatment efforts, none of which seemed to have a therapeutic effect, and so Gerald clearly qualified to be classified as "treatment resistant." Finally Gerald was considered to be a candidate for electroshock treatment, but at the last minute the new psychiatrist to whom the patient was referred suggested a trial of psychotherapy.

Gerald was eager to try psychotherapy and began with a good deal of enthusiasm and soon realized that merely talking to someone made him feel much better.

Gerald's past history was significant in that he was an only and lonely child who recalled that his present-day feelings of his supposed clinical depression were replicas of his lonely childhood. He would wait for his father's late return from work and sit with him as his father ate dinner for a brief period before having to go to bed. These memories were of yearning to be with someone to relieve his loneliness. Gerald likened his early relationship to a rather taciturn father to that of the psychiatrist who sat with his prescription pad and other therapeutic modalities with but little or no time for listening. Gerald's transference to this psychiatrist was readily apparent to the analyst therapist, and without going into the details of the treatment the ensuing psychoanalytic psychotherapy had a successful outcome. When Gerald also noted that his adult life was punctuated by intense and painful periods of loneliness when he was away from his wife for any reason, the therapist concluded that the diagnosis of a prolonged mourning reaction manifesting clinical depression was probably a mistaken diagnosis. The distinction between the overt phenomena of mourning and that of intense loneliness is certainly an interesting theoretical problem, but the crucial element once again rests upon the emerging transferences that develop. For some patients this is the essential element for the proper prescription of treatment.

Distinguishing psychoanalytic explanations from psychoanalytic pathology

The initial popularity of psychoanalysis and its innovative insight into the workings of the mind has led to its employment in a variety of arenas ranging from literature to history and on to medicine. New ways of investigation and explanation were brought into topics, and they seem to bring new life to many of what may otherwise be seen as dormant studies. Hamlet came to be seen as someone locked into an obsessive compulsive struggle, Julius Caesar as a player in an Oedipal struggle and Adolf Hitler as having a borderline personality disorder. The division between explanation and psychopathology became blurred, and this blurring is seen in many of our textbooks (McWilliams, 1994) as well as in our characterizations of some of our political leaders (Weston, 2008).

It is important to recognize a distinction that psychoanalysis offers as it hopes to explain certain phenomena without pinning labels of sickness on what is explained. For instance, it may be of interest and revealing to see the relationship between Othello and Iago as indicative of a homosexual interplay without necessarily resorting to a DSM exercise in labeling. Napoleon was surely compensating for his small stature in his exercise of power, but that need not mean he had a narcissistic personality disorder. Explanations of Hitler's anti-Semitism, Bach's care in balancing his toccata and fugues, Obama's preoccupation with hearing all sides of controversy and on and on are all fascinating speculations that may employ psychoanalytic ideas and formulations, but none of these should be confused with diagnosing mental illness from afar. The data for that are not available. Of course there are many other ethical and moral reasons for not engaging in such speculations, and there are likewise many occasions to make psychiatric diagnoses on persons of notoriety. The distinction to be made is one of using psychoanalysis to expand our vision about many areas of interest versus making psychoanalytic diagnoses about persons who are not in analysis and so cannot offer the data of psychoanalysis to us. We can guess about their diagnoses, but then we are running a risk of comparing persons with similar behavioral phenomena. Such risks are not uncommon, but we should keep the distinctions in mind.

And psychodynamic formulations

In the evaluation of patients for psychotherapy or psychoanalysis or any of a variety of therapeutic interventions, it is common to prepare a psychodynamic formulation of the patient's illness. This formulation is constructed from the patient's history, the presenting complaint and the material gleaned from the interaction between patient and therapist during the diagnostic sessions. It serves as a guideline or roadmap both for disposition as well as for the further pursuit of therapy. It is probably fair to say that one's initial formulation turns out to be sometimes right and sometimes wrong. Patients who may fit one diagnostic formulation may well surprise or even startle us. Here may be one problem with the dynamic formulation: it can lock us into a set of presumptions that restrict a certain freedom of inquiry. Psychoanalytic work often requires an openness to surprise, and it is possible that diagnosis may inhibit or foreclose that.

Discussion

It may well be a purist exercise to insist on utilizing psychoanalytic data to diagnose mental illness, but at the onset this purism is but to restrict *psychoanalytic diagnoses* to conclusions drawn from such data. The danger in going outside of such data is one of reducing the study of mental illness to mere matters of opinion, and surely everyone has a right to their own opinion. We have seen that opinions are close relatives of prejudices, and so we drift farther and farther from a hoped-for scientific stance. Although psychoanalysis lacks the hard objective evidence that many scientific pursuits enjoy, our theories do direct and constrain what we gather as evidence, just as the study and examination of our countertransference limit our personal preferences. The personal analysis of our analysts is another device to direct us to greater objectivity. With all of the pitfalls and shortcomings of the analytic process having a claim to a scientific status, we are only more unlikely to achieve such a status by joining our psychiatric colleagues in the pursuit of diagnosis by plebiscite.

Conclusions

As long as we claim that transferences are ubiquitous and universal we may find ourselves trapped in the vise of our proposed pathology's having the same lack of clear delineation. What is needed is an evaluation of transference that is not based on those moral issues that allow the "oughts" to be equated with likeability or something akin to the consensus view of psychiatry. In other words, are there transferences that are good like a "good" toaster versus those that are bad? We do seem to agree that a successful analysis need not result in the disappearance or dissolution of transferences, but we instead offer the somewhat vague concept of "resolution." Somewhere between the statement attributed to Lawrence Kubie that a successful analysis should end with the analysand even forgetting the analyst's name – i.e., all transference is eliminated – and the equally troublesome ideas that all analysis does is akin to rearranging some sand on a beach (attributed to Hans Sachs), there should be a psychoanalytic distinction that might enable one to distinguish the pleasure in gambling from pathological gambling, sexual perversion from sexual preference, etc.

What I offer has, not surprisingly, a number of qualifying phrases that range from "for the most part" to similar needs not to be pinned down too severely. It is the aforementioned distinction between agency and host, the difference between "what I am" and "what I do," from "what I harbor" and "what is done to me." Agency is the crucial ingredient in normativity (Caston, 2011). There are many theoretical words and concepts to embrace the concept of agency ranging from executive function to self-control or self-actualization to variations on the concept of consciousness. Another way to characterize the basis of psychoanalytic thinking is the move from the subpersonal best representative of neurophysiology to the personal of our psychology. No doubt psychoanalysis has a good deal of work to do to better present our particular notion of the normal. But that is a much more desirable position than to merely claim that it is beyond our science. The data of psychoanalysis must be taken as representative of its status of normality versus pathology. We cannot dodge that responsibility. Inasmuch as this chapter does not and cannot do more than point the way for a psychoanalytic foundation for assessing normality and pathology, the task ahead is for our own DSM to be formulated.

Summary

Psychoanalysis approaches the diagnosis of mental illness differently than does psychiatry, which determines categories of psychopathology by phenomenology with a hoped-for more scientific formulation based upon neurosciences. Psychoanalysis gathers the information needed for analytic diagnoses from the analytic process and not from behavioral descriptions or brain studies. We often share names for disorders with psychiatry, but these shorthand pursuits should not seduce us into a claim for an analytic diagnosis. Thus we may agree on a diagnosis of neurosis or psychosis, but only a neurotic or psychotic transference can allow for an accurate psychoanalytic diagnosis.

The confusion that has grown up over sexual preferences and their membership in the categories of psychopathology has resulted in unwarranted pain and discomfort because analysis has failed to insist upon analytic data to make conclusions about what should be called abnormal. Norms are not to be based upon personal ideas about right or wrong either in morals or sexuality. Our science must be our guide.

Part II

The newer models of the mind and the self

Introduction to Part II

In order to better position psychoanalysis as a study of the mind and the self while thoroughly aware of its dependence on the brain, it is necessary to distinguish the mind as a separate area of inquiry, and the same for the self. Just as the expert violin maker is able to appreciate a virtuoso violinist without reducing the performance to the instrument, one must employ a working theory of the mind as the study of meaning and the self as the agency of mental life. The first demands a different theory of the mind than that of primarily and exclusively a product of the brain. The Kurzweil (2012) book, which collapses the brain and the mind, is a good example of an exercise that exemplifies the violin maker's ignoring the music of the composer. This new theory is one of an expanded theory of the mind. The delineation of the self requires a better model of our connection to the world than that of an internal replica that we refer to in order to stabilize our position in the world along with our relationship to others. The latter is not accomplished by a cognitive standard of comparison – i.e., "George looks angry because my internal model of the angry George is activated" – but rather by our exercising our skill of empathy, which allows us access to the world and to others. As Nöe says of Merleau-Ponty:

> [E]mpty heads are turned to the world. The world is not a construction of the brain, nor is it a product of our own conscious efforts. It is there for us, we are here in it. The conscious mind is not inside us, it is, . . . a kind of active attunement to the world, an achieved integration. It is the world itself, all around, that fixes the nature of conscious experience.
>
> (Nöe, 2009, p. 142)

Chapter 6

Being kept in mind

We are all familiar with the phrase "being in mind," with all of its many variations ranging from wishing to be "kept in mind" to "being thought about" to "being lost from mind" or "dropped from mind" and its own unfortunate equivalents amounting to "being forgotten." No doubt it is not peculiar to those who treat the mentally ill that there are patients who we seem unable to *not* think or even worry about, and so seem ever to be "on our minds." Indeed, though it may be more common with potentially suicidal patients, there are those who seem to have carved out a permanent or semipermanent place in our minds, so that we seem never to be free of our concern about them. They worry us or even delight us.

At a certain moment in the treatment of our worrisome patients we should, or need to, recognize that it is important or even vital for such patients to be thought about, and so having a place in our minds may require this negative, irritating or enjoyable quality. However, that is best thought of as the price of such residency. The more fortunate position would be that of a pleasurable presence, but there is little doubt that worry often trumps a fond memory.

Once we recognize the intense need not to be forgotten, even at the price of continual annoyance, we are able to evaluate the need both diagnostically and developmentally and so consider a proper therapeutic approach. This may require an alteration in our usual conceptualization of the mind in terms of both its formation as well as its scope, and this is what will be proposed in this chapter.

Development

As noted earlier, Piaget (1973), in his study of child development, found that children, like so-called primitive people, did not know or believe that the mind was confined to the head but rather felt that it extended to the world around them. By about the age of 11, most children (like most neuroscientists) arrived at the "correct" view that images and thoughts are situated in the head. As long as the mind is identified with the brain, the model that is employed involves the transposition of the furniture of the world from the outside to the inside. Psychoanalysis does this by utilizing the concept of "representations," which, in a rather crude way, consists of miniaturizing and moving the significant person or objects of our lives to a place inside of our heads. This is a convenient and quite workable model that should be seen as just that – i.e., a model.

In contrast to the theory of internalization and registration is a model of a mind that extends into the world around us. This is the extended theory of the mind, and it considers the mind as going beyond the head and so treats meaning as lying in a web of activity with the environment. This is an alternate or open model of mind that dispenses with the concept of internal representations. It may be prudent therefore in thinking of the mind primarily as an activity rather than as composed of mere printouts of the brain. Self psychology is one model that employs this extended theory by its employment of other persons as selfobjects that constitute the self. That is one example of the utility of this model.

Diagnosis and psychopathology

We earlier wrote about the Hollywood movie *Home Alone*, in which a child is left behind when the remainder of his family goes off on a trip. He is literally forgotten or lost from their minds. The child, a boy, appeared to take for granted that he would be thought of by the family inasmuch as he was convinced that he was "in their minds." It is not uncommon for a child in a group setting such as a picnic or at a beach to wander off. The child is in the so-called primitive stage noted by Piaget in his belief that he is ever in the minds of others.

For the most part the essential enlistment of others in any form of a requirement of being thought about is felt to be correlated with the more extreme forms of pathology. Gerald Adler has written extensively on this problem, which he insists is pathognomonic of what he calls borderline psychopathology. He describes such patients as fearful of being alone and unable to maintain positive images of sustaining people in their present or past lives (Adler, 1977, 1988). He proceeds to offer a theoretical understanding of aloneness by utilizing Piaget's stages in the sensorimotor development of an object concept. In brief, the problem lies in the inability to reach stage VI in which the child possesses a sustained mental representation of the object. He describes such patients as struggling with feelings of emptiness, of needing to phone to hear our voice, of being unable to effectively remember us, of using an unpaid bill or an appointment card to help them evoke positive memories of us and of requiring a postcard from us while we are on vacation to help them remember us just as they wish to be remembered by us. They cannot keep us in mind, but alongside this, they need for us to keep them in our minds.

Some clinical illustrations

A psychiatric resident presented a suicidal patient to a conference with the accompanying commentary that he was continually worried about the possibility of the patient's suicide and at one point drove to the patient's house to bring him to the hospital inpatient service. He was soundly criticized for his behavior and was told that he was being manipulated by the patient.

An analytic patient called to ask for an extra appointment with the avowed reason of making up for an earlier missed appointment yet with no evidence of needing the appointment. The patient said that the regularity and predictability of the schedule offered a feeling of safety and solidarity that had nothing whatsoever to do with the content of the material discussed. The analyst was happy to offer the supposed make-up session, but found it to be barren of new material. There were no fantasies about the supposed missed hour and indeed none of the previous missed hours revealed any significant material about the activity of the analyst. Once again the crucial issue seemed to be that of regularity.

Another analytic patient was exquisitely attuned to future disruptions in the schedule, whether they be due to national holidays or vacations or unavoidable appointments by either the patient or the analyst. The anticipation did seem to lessen the impact of the miss. The planning of the dreaded disruption offered a modicum of control over what was basically uncontrollable.

Inasmuch as Adler believes these patients lack a positively enduring memory of others, he recommends treating them with frequent face-to-face sessions and allowing telephone calls and the above-mentioned postcard while the therapist is on vacation. He makes little mention of the fact that this cognitive deficit seems to simultaneously impact the therapist as well, other than to claim that projective identification of the "commitment self or object representation" may occur (Adler, 1988, p. 3). Thus the ever-present thinking or even worrying about a patient is a result of the patient's projection of his or her problem into the therapist's mind. The psychiatric resident who picked up and drove his patient to the hospital had the patient's anxiety projected into his mind. The patient who anticipated a holiday or a vacation projected this future event into the therapist's mind. Of course we can say that the therapist was empathic, or identified, with the patient and so understood the patient, and in that manner we need not have the patient physically present in order to feel as he/she does. Yet another possibility presents itself in the need of the patient to be thought about; i.e., it is not merely the achievement of evocative memory that characterizes the capacity of a person to feel in possession of sustainable, internalized object representations, but it is also the real existence of said objects. Living with others who cannot think about you not only disallows what Fraiberg and Adler call "person permanence," but sends one on a lifelong voyage to find others who, by thinking about you, allow for your very existence (Adler, 1977).

An alternative theoretical model

Enactivism is a theoretical approach to understanding the mind. It emphasizes how the human mind organizes itself by interacting with the environment. This approach criticizes representational views of the mind and emphasizes embodiment and action. As Maturana and Varela have said,

> Our existence is coupled to the surrounding world which appears filled with regularities that are at every instant the result of our biological and social

histories.... The whole mechanism of generating ourselves is, as describers and observers tell us, that our world as the world we bring forth in our coexistence with others will always have precisely that mixture of regularity and mutability so typical of human experience.

(1992, p. 255)

This states the basic tenet of our mind, which is actively organizing itself on the regularities and predictabilities of the world.

This model also states that our minds are not confined to the brain in our head but that they extend through time and space. It is not only the sad fate of the borderline patient that he is so incapable of retaining a positive and lasting image of others but in fact the fate of all of us that we rely on the reality that we are thought about by others. Surely our patients need us to various degrees and in various ways, but we as therapists organize ourselves with our patients who ultimately take up residence in our minds and thereupon become components of our minds.

It goes both ways

In a somewhat critical review of a book that made the claim that borderline patients could and should be treated in a classical ego psychological manner, Adler insists that these analyses missed some crucial aspects of their patients' psychopathology, especially in terms of issues of abandonment and aloneness. Of course there is no clear-cut explanation for the fact that patients may be seen differently and treated differently by analysts of varying backgrounds, training and personalities. Inasmuch as the diagnosis of borderline pathology relies a good deal on both objective factors, such as a developmental history, and subjective factors, such as those attributed to countertransference and projected identification, it should come as no surprise that different analysts often disagree. The crucial element in this particular disagreement has to do with the activity of the therapist. Adler feels that telephone calls are often required, that these patients cannot tolerate silences and that a postcard is needed while the therapist is on vacation. He does temper these recommendations with the fact that such active interventions will ultimately be interpreted and therefore dispensed with (Adler, 1988, p. 370). The book that Adler is somewhat critically reviewing, however, describes the analysis of their patients as being no different from that of any neurotic patient and makes no allowances for such activity on the part of the analyst. These supposed needs of the patient are either nonexistent, or they lie in the mind of the analyst or therapist.

This raises the interesting question of whether we find our patients or whether we construct our patients. No matter which, we surely need our patients. Although our diagnostic categories are often posited on the particulars of how our patients may need us, they are rarely if ever configured on the principles of our needs for them.

Case example

Dr. F. was referred for analytic treatment by his medicating psychiatrist, who worried about Dr. F.'s getting into trouble because of multiple sexual affairs with his patients. Dr. F. was an internist with a large practice along with an unhappy marriage, which had led him into a number of efforts at psychotherapy. He had left a previous analyst who had suggested that F. was upset at not seeing this analyst during a vacation. Dr. F. felt this was an outrageous idea – i.e., his thinking about and missing the analyst. In his new treatment he decided to seek a divorce from his wife and proceeded to effect a separation as well as to reduce the number of sessions of treatment primarily because of financial pressure. He had earlier noted that he was getting a lot out of his therapy, and so he regretted having to cut down. The therapist felt alternately pleased and disappointed.

Rather than a report on the conduct of this treatment, the focus here will be on the recognition and meaning of the mutual needs of the patient and the therapist. In Dr. F.'s childhood he was the star of the family and the only one who could make his chronically discontented mother happy. He had no conscious need of his father, who was without ambition and who rarely took a position on any topic. Dr. F. was himself quite important to his many patients and confessed to overprescribing pain prevention medication as well as antibiotics. He liked being needed and was constantly being called by his patients for a variety of petty complaints.

After more time in therapy, the patient and his wife separated, and the patient took up with a new woman companion. He was quite happy, save for the relationship with his teenage son, who would have no more to do with him. It was so painful for Dr. F. not to be needed by his son that he contemplated suicide. It was not clear if the patient needed his son, or if he needed to be needed or if such a distinction is possible. In the treatment the patient could be seen as needing the treatment but unable to acknowledge such a need. Any interpretation of the patient missing the therapist was regularly scoffed at, all the while the patient making certain that substitute hours were available for future holidays. Dr. F. had to be needed but had great difficulty in an acknowledgment of his needing others. The therapist could recognize a variety of feelings about his patient, but could not admit to needing him.

The therapist of Dr. F. felt pride at a possible resolution of a very difficult marriage and especially over the comment of Dr. F. that he wished that he could come more often. The treatment seems quite representative of the mutuality of needs that characterize all treatments. The analyst or therapist is not the repository for the internalization of the internal world of the patient any more than the patient serves as the site of the analyst or therapist's projections. In a sense they each need the other and so take up a place in each other's minds.

It is the thesis of Gerald Adler that borderline patients need us in a particular manner because their pathology reflects severe anxiety over abandonment and aloneness. A modification of this thesis is that all patients need us, and all therapists and analysts need their patients for different reasons and with different intensities. Thus there exists a mutuality of needing. This is best explained by seeing our minds as extending into and constructed by the environment rather than composed of a set of internalized representations of objects. That latter position leads to the analyst or therapist being the blank screen for projected representations. The extended mind allows for variable meanings dependent on the contribution of both patient and therapist; i.e., we need each other.

The mind in this alternate model is a self-organizing activity that interacts with the environment. The therapist or analyst is thereby included in this organizing activity. Rather than the patient projecting an image of an unconcerned parent onto the therapist, the therapist is enlisted as a part of a fragile self that is ever in search of an unachievable stability. At this point it seems to make little difference as to the choice of models. However, if one thinks of oneself as a vital participant in the construction of the patient's psyche, then a different therapeutic activity may suggest itself. The interpretation to be offered necessarily involves the experience of the therapist but is not based on a form of accusation – i.e., "you are making me feel" – but more in the form of an empathic understanding – "I know how you must feel." Yet that may still allow for little distinction between the models. However, one could support Adler's actions by offering to become a more positive and helpful component of the patient's mind. Indeed this stance does support certain relational interventions; i.e., one corrects or ameliorates the patient's deficits, which are revealed by way of the experience of the therapist.

For the most part, one can readily distinguish a position or feeling that is foreign to oneself from a projection from an authentic experience. Indeed, the very idea of having something projected into someone else is usually considered to be outside of one's person. One may be accused of being in a particular emotional state, but that is readily recognized as not belonging to the self. Seeing oneself as part of the other is a more easily acceptable concept. Such a temporary identification is the basis of empathy and so is the way we understand another person.

Discussion

We may organize the presence of thinking about patients on a continuum from occasional to extreme. Some patients are thought about when writing notes or going over one's schedule. Some stimulate an almost relentless and unrelieved preoccupation. These extremes are regularly thought of as reflecting pathology and so range from the mild neurotic to the severe borderline. This does not always hold true, since some patients that are not particularly troubled may delight and please us and so may take up a good deal of our private musings. We may even feel guilty about enjoying these patients while wishing to avoid that troubling group. Being preoccupied with thoughts of a patient can also take the form of a positive activity, much like falling in love with a patient and not being able to get him/her out of one's mind. Many boundary violations are pathological expressions of various forms of lovesickness or an obsessive focus on a patient.

From the extreme of a patient's need to be thought about by a therapist to the opposite pole of a therapist's total obsession with a patient, we see the pathological forms of an often unwelcome intrusion into one's mind. Essentially, the subscript to this unwarranted presence is the plea "pay attention," and we are called upon to do so from the real needs of a patient to our own needs to follow that instruction. It is difficult to equate these needs, but we feel it easier to claim the patient's fear of abandonment as leading to our keeping them in mind and worrying about them; whereas it is more difficult to see our need to matter as the corollary to enjoying thoughts of another.

One may of course utilize a number of possible psychodynamic explanations to the problems of excessive worry to lovesickness, but the point I wish to make is that of the mind extending out to and including the world as opposed to that of a locked-in brain looking out at and representing the world. As Alva Nöe says, "we are out of our heads. We are in the world and of it" (Nöe, 2009, p. 183). The operative word in that quote and in the title of this chapter is "in," which unfortunately too often means "inside" and then conjures up an "outside." Our world is not bounded. It is all of us.

Summary

At times we cannot get someone, often a patient, out of our minds. There is one possible explanation for this phenomenon that is attributable to the patient and another explanation that is attributable to the therapist. Worrisome patients are frequently described as those who fear aloneness and abandonment and so in some manner cause us to be available to them. Patients who are especially pleasing to us preoccupy us because of our personal needs, which may be of more significance than that of the patient's, yet still allow for the assignment of some causal factors to the patient.

Apart from the extremes of worrying about patients, making ourselves unusually available to such individuals and obsessing about patients who become almost necessary for our well-being, there is a large middle ground of our needs to be both thinking of and thought about. Each perspective demands a mutual contribution of both patient and therapist and so asks for a model of the mind that does not depend upon the patient's projecting a set of thoughts and feelings into the mind of the therapist. Rather, a model that is ever open to the environment suggests an open mind that organizes itself with the contributions of the surroundings. Utilizing such an open model may contribute to differing diagnostic assessments that are arrived at by analysts of differing theoretical persuasions. The hoped-for objectivity of a diagnostic evaluation is tempered by the personal need and prejudices of the evaluator. Although borderline patients may be more likely to stimulate most of us to become more available to them, and glamorous patients may make us more likely to devote extra time to thinking about them, some of us may be immune to either activity. There is no doubt that patients impact us, but our availability or our openness to them may well be the crucial factor. Now to see how we gain access to the mind of another.

Chapter 7

The many meanings of empathy

The intent of this chapter is twofold: the first is that of an effort to organize and collate what seems to be a rather diverse collection of definitions of empathy and the second is to make an effort to provide a specific definition of empathy that is especially applicable to psychoanalysis. Within the first effort there will be a necessity to better integrate the contributions of neuroscience along with the role of oxytocin, while within the second the manner in which empathy plays a unique role in a hermeneutic science needs to be clarified. The background to this dual effort is the recognition that empathy develops over time and this development may or may not be evidenced in the clinical setting. Thus when a patient enters with a cheerful face or a stricken demeanor, we may immediately read that expression, all the while knowing that that particular reading may be then elaborated over time. The initial empathic grasp is something that can be seen in infants and children, while the more detailed understanding is a developmental achievement. Each and every interpretation allows for a new and variable way of understanding, and so our empathic efforts must be sustained over time. Development should serve to clarify the many varied definitions of empathy, while sustained empathy may be needed to better situate empathy and its role in an interpretive science.

Introduction

In most of the discussions about empathy, a disclaimer is routinely introduced in terms of the fuzziness of the concept, along with the fact that there are many definitions of the word. This plethora of definitions is explained by an implicit agreement that everyone is entitled to an opinion, and this is an arena of opinion. Usually the author(s) proceeds to offer his or her own definition of empathy without any sort of acknowledgment of it being at odds with other definitions or any effort directed at justifying why his/hers is the most appropriate. One representative article presents empathy as composed of three skills and proceeds to select that of emotional contagion as one such skill with no recognition that other authors consider it to be empathy in its entirety based upon what is called "matched neural representation" (Preston & de Waal, 2002). This is the definition relevant to empathy in animals, and for that usage it is sufficient. Whether it can be extended beyond that is not addressed.

It is of course difficult to reconcile the existence of empathy in animals with the rather complex set of skills that some authors insist do compose empathy. Batson (2009) proposes a set of eight "related but distinct phenomena" as being the ingredients of empathy and in doing so concludes that empathy is not one single thing. Of course there are many terms and concepts that are not one single thing but rather belong to a family of common or joined meanings. One way to understand and investigate such commonality is to study the growth and development of the concept. We study the child's development of locomotion from crawling to walking to running as well as a host of other developmental lines that qualify as "distinct but related" in such a manner. So, too, can one study only one phase of the development in its own right without dismissing the significance of the others. Empathy appears to be composed of a complex collection of ideas and abilities that develop over time.

I propose that we consider empathy as a family of concepts that emerge in sequence during normal development but that follow the Glover recommendation that

> what we call in the structural organization of mind is not to be thought of merely as a series of superimposed developmental levels. There is, to use a spatial image, a vertical as well as a horizontal development of the apparatus.
> (Glover, 1950, p. 374)

Therefore, empathy should best be organized as a process moving from, say, emotional contagion to simulation to mind reading to what a majority of investigators would term mature or generative empathy, and then on to sustained empathy. Each of these phases may or may not incorporate earlier stages yet may develop in its own right. Thus simulation or mimicry is regularly seen in the form of somewhat sophisticated imitation. So, too, perhaps a different perspective may be offered to such concepts as projective identification, which may be seen as a particular developmental moment in the line of empathy. What appears as development is never one-dimensional, nor need it be linear.

Of course the developmental line of empathy is primarily offered as a psychoanalytic contribution in order to bring order to the many uses of the term and so better position it in the practice of psychoanalysis. If elephants and little children are said to be empathic (de Waal, 2009), and if poker players and morticians are exemplified as outstanding examples of empathic understanding, then it seems quite crucial that psychoanalysis makes it clear what empathy means both to the field as well as in the practice of its discipline.

At crucial points in the line of development we will need to determine if we have entered a new level of comprehension that demands a new vocabulary along with a new level of meaning and understanding. Initially, one must consider these levels of meaning.

The continuum vs. the magic moment

If one were to study the development of any complex system that was derived from simpler elements there is usually a point where a change in comprehension is called for. The study of letters forming words and of words forming sentences and of these sentences conveying meaning is easily and readily placed on a continuum, yet the shifts in what we have been terming "universe of discourse" demand that we recognize that an entirely new form or level of understanding is required. These "magic moments" are subtle in some cases and dramatic in others, and there is an ever-present urge to attempt to capture and dissect these moments. Of course we may never be able to make sense of the transformation of the television pixels to the drama unfolding on the screen, but the effort to better capture the neurons of the brain becoming the emotions and ideas of the mind does today dominate a good deal of time and effort. If an area of the brain is indeed the site of anxiety or depression, then surely one can honestly say that "the brain feels" or "the brain wants" or posit any other personification of the brain. Just as we naturally say that "my stomach doesn't like spicy food," can we not talk of organs as persons? If we do speak of our personal parts in this manner, we recognize that these attributions are conveniences of speech rather than realities of anatomy. However, we seem not to make this distinction with the brain and ever hope that some careful study of those neurons will reveal the magic moment of thought and feeling.

One good example of our preoccupation with the continuum can be seen in the study of empathy. It is something that can be considered both in its neurologic origins as well as in its psychological development. The temptation to construct a continuum is apparent. The start of the continuum begins with the earlier noted neural basis for imitation. This was when Giacomo Rizzolatti and his colleagues discovered neurons in the premotor cortex of monkeys; when one monkey picked up a peanut, these neurons became active when seen by another monkey (Rizzolatti et al., 1996). The possible culmination of the continuum is nicely expressed by Roy Schafer's definition of empathy: "empathy involves the inner experience of sharing in and comprehending the momentary psychological state of another person" (Schafer, 1959). If we compare that to the "matched neural representation" (Preston & de Waal, 2002), which is the definition offered as applicable to animals, we can see the sharp distinction between the psychological and the neurological. However, if we next compare the Schafer definition to that of Heinz Kohut, who says that "empathy is the capacity to think and feel oneself into the inner life of another person" (Kohut, 1984, p. 82), we may begin to initiate a definition that is primarily psychoanalytic; i.e., a step beyond the mere psychological.

The pursuit of the "magic moment" is often elusive, and the difficulty in capturing it is often a result of its unattainability; i.e., there may be no such point of transformation. Oliver Sacks, in his rather comprehensive survey of hallucinations (Sacks, 2012), devotes a good deal of his study of the phenomenon to pinpointing the brain areas involved while also cautioning us that "the power of hallucinations is *only to be understood* from first person accounts" (Sacks, 2012, p. xiv).

Once again, a "person" or the concept of a person is needed to fully explain. Finally, in a chapter called "The Haunted Mind" (pp. 229–254), Sacks seems to require the mind to "experience" the brain. That chapter is one of accounts of individuals who have or see or experience hallucinations and reads as that of minds that exist outside of and separate from the earlier described brains. However, this is not an effort to dismiss the brain as much as one to aid us in the comprehension of phenomena that exist in separate universes of discourse. For Sacks it is not a continuum.

If we return to the study of empathy we are called upon to add another dimension – that of the development of empathy; i.e., as we move from the neurologic "matched neural representations" to "the inner life of another" we must reckon with the purely psychological changes in the concept, and so we must consider if the meaning of empathy changes as well.

The question that presents itself is whether "matched neural representations" are but a step on the way to "the capacity to think and feel oneself into the inner life of another person" or if it is better to consider it as a necessary part or component of the latter. There is a difference. Please forgive this detour to a rather common conundrum in philosophy sometimes thought of as the Chinese Room Problem. In brief it describes a closed room in which a person receives Chinese symbols and proceeds to look them up in a book that translates each symbol into English. The person then sends the correct translation of the symbols out and so in effect serves as a translator of Chinese into English. The question that is thereby offered is whether the person understands Chinese or if such understanding belongs to the room or to the entire setup or indeed if anyone or anything understands Chinese. This, of course, is not much different than groups of letters forming sentences or groups of neurons giving rise to meanings or matched neural representations "giving rise to empathy." Understanding is a mental operation that is absent from the Chinese Room.

Oxytocin, empathy and discourse

Oxytocin is a hormone manufactured in and by the hypothalamus and released into the bloodstream by the pituitary gland. It is said to signal the uterus during childbirth as well as to stimulate the release of milk for nursing. In work with rats and sheep it appears to enhance mother-infant bonding and in further work with prairie voles it fostered pair bonding. Oxytocin and a related hormone called vasopressin are said to play a role in the modulation of social and reproductive behavior (Mlodinov, 2012, p. 94) and, not surprisingly, Oxytocin was tested on humans in studies ranging from the treatment of autism to borderline personality disorders (Miller, 2013, p. 264). It seemed to increase trust in some subjects while making others less trusting. A nasal spray of oxytocin allowed some to "read what's on someone's mind from the look in their eyes;" i.e., it was felt to be essential for empathy. However, more testing moved researchers to decide that "it depends on the situation in which it is given or the person to whom it is given." The popularity of the drug is primarily being encouraged for the treatment of autism, and trials are now going on for its continual use for weeks or months. Yet researchers seem to have concluded that it is a "double-edged sword promoting bonds with familiar individuals but promoting unfriendly behavior toward strangers" (Miller, 2013). These conclusions are ignored or unrecognized in popular books that claim universal or general effects for oxytocin (Panksepp, 1988). The problem is best resolved with the realization that our engagement with the world is not based on a mere repositioning of the environment into our head, or what is called the "internalist approach." Rather we are ever involved in the world with our expectations and our shared meaning with what is termed the "externalist approach." Inasmuch as we are always participating in the environment, and we are also the products of a personal history of growth and development, it is no surprise that each individual is reacting to a drug with a somewhat unique response. As Nöe has said,

> Experience is not caused by and realized in the brain, although it depends causally on the brain. Experience is realized in the active life of the skillful animal, i.e., the person and his or her life. A neuroscience of perceptual consciousness must be an active neuroscience – that is, a neuroscience of embodied activity, rather than a neuroscience of brain activity.
> (Nöe, 2004, p. 227)

We can repeat what Paul Ricoeur has further said:

> I profess considerable skepticism with regard to the possibility of constituting an overarching discourse . . . above and beyond the profound unity of what appears to me sometimes as a neuronal system, sometimes as a mental experience. In the last analysis we are dealing with two discourses of the body.
> (Changeux & Ricoeur, 2000, p. 294)

Ricoeur also insists that "we can understand either a mental discourse or a neuronal discourse, but . . . their relation to each other remains a problem because we have not managed to locate the link between the two within one or the other" (Changeaux & Ricoeur, 2000, p. 69).

Nöe and Ricoeur are articulating a point of view that says that even if "matched neural representations" are an essential part of what psychology and psychoanalysis consider to be empathy, they are not a part of our mental discourse, which involves beliefs, desires and wishes. Indeed there is no way of uniting the two languages. There is no magic moment of transformation. We merely move into a different form of understanding.

The developmental line

Although a time line is a fundamental tenet of development, we need to exercise caution in terms of the appearance and persistence of the various components of empathy, and we must keep in mind that earlier phases are not replaced but may continue to develop in their own right.

1   *Emotional contagion*
    There is no doubt that people are able to feel themselves into another's emotions, just as some animals are capable of doing and as is observed in infants. There is even an Emotional Contagion Scale that has been translated into a number of languages (Hatfield et al., 2009, p. 20). Many researchers feel that either this alone is empathy or it may be considered a precursor of empathy. Emotional contagion is said to be the basis of mimicking and is often divided into facial, vocal and postural mimicry. Studies initially by Darwin and later by Ekman et al. (1983) feel that facial muscles are the basic component in the transmission of emotions. However, people can also "catch" emotions, as seen in infants, music lovers and religious revivals. Thus, one can begin the developmental line with emotional exchange with care that the line need not be seen as one unbroken continuum.
    The linear study of empathy is nicely seen to begin in reports that indicate that one-year-old infants understand that, just like their own actions, other people's actions have goals (Decity & Meyer, 2008). By two years of age children display the fundamental behavior of empathy by having an emotional response that corresponds with another person (Hoffman, 2000). During the second year toddlers will play games of falsehood or "pretend" in an effort to fool others, and this requires that the child knows what others believe before he or she can manipulate their beliefs. This is mainly cognitive. Children usually become capable of passing "fake belief" tasks considered to be a test for a theory of mind around the age of four (Baron-Cohen, 1995).

2   *Mind reading*
    Mind reading, along with a host of related terms such as mentalization, belong to a category of cognition that too is often collapsed as being equivalent to empathy. However, it may be useful to see its growth as at times being independent and at times as combined with affect. Michael Lewis reviews the development of "knowing" in children (Lewis, 1993). He lists four levels of what he terms the "knowing," which prevails from birth to 18 months, followed by the achievement of knowing what others know, up to and including knowing that others know what you know. Many cognitive psychologists have contributed to the study of both how we read our own mind by way of introspection and how we read others' minds (Carruthers, 2009). At some point in the descriptions and discussions of these topics mention is made of the emotion associated with this knowledge, and often there is a switch in the vocabulary from knowing to empathy. Imagine the following scene of a person descending in an elevator with a glass door and seeing a woman holding a child's hand with her

head down. She is preparing to enter the elevator as soon as it descends and the door opens, but a man inside the elevator sees her and concludes that she will walk right into him unless he alerts her. Now he is clearly reading her mind, but the question arises as to whether he is empathic with her. If he were to know that she was on her way to visit her sick husband in the hospital, he would know more and perhaps be able to "feel" along with his cognition. At least in theory we are able to isolate the knowledge from the affect, and to insist that empathy needs both. One may call both scenarios as indicative of empathy, but they are different appraisals and do need to be distinguished. Indeed some distinguish cognitive empathy from emotional empathy.

3   *Location and perspective*

Locating what is known along with what is felt together brings about a whole new level of complications and possible confusions. We must begin with a number of categories. As a start we distinguish, with our earlier distinction, between knowing how someone feels and feeling how someone else feels. We then move to putting yourself in another person's place or mind or situation. This corresponds to Lipps's (1903) original state of "Einfuhling," which was later called empathy. This position can be called projection or perhaps imagining how someone else feels. Once in that place we may literally feel that person's affects such as distress or happiness. At this point one must construct a split between a complete identification with that person and a capacity to "think about" how someone else feels. This may well be the problem with projective identification, which seems to equate the beginning achievement of empathy with something akin to an accusation in the form of the other person or patient's being responsible for your state of mind. Here the combination of cognition, emotional contagion and what is called meta-representation is to be considered as a further and necessary developmental step. Once again this particular combination is considered to be, and is called, empathy by many.

Carruthers feels that mind reading precedes our considering our own minds as in introspection (Carruthers, 2009). However, regardless of that, it is important to differentiate knowing about oneself, knowing that others know about you and knowing that you know that others know about you. For empathy, one must know and feel that you know and feel how someone else knows and feels. Knowing is not enough. Feeling is not enough. The combination must also be represented and so considered as apart from oneself.

This developmental line of empathy can be said to reach an endpoint when one can contemplate how one thinks and feels about how someone thinks and feels about someone else. It begins with feelings, adds cognition and then frames it to examine it. No doubt a number of other additional stages or phases of empathy can be added to this line, which is presented here only in outline.

Clinical considerations

The usual expectation in psychoanalysis is that the analyst strives to be empathic with his or her patient in order for the patient to ultimately understand or be empathic with himself or herself. However, just as we confront patients at their various developmental levels in terms of defenses, object relations, levels of reality, etc., so too must we be prepared to consider and evaluate them at their level of empathic development. Likewise, it is assumed that empathy serves as a source of data gathering as well as in performing a therapeutic function in its own right.

Complications ensue with the recognition that the combination of knowing and feeling, often described as "understanding" another, regularly becomes linked with what one does with this understanding. There is an understanding that stands alone and can be considered to be a mere reading of another person, along with an understanding that is coupled with an action – i.e., doing something with this reading or comprehension. In the position of caring for another we may make efforts to feel for another and so ameliorate a painful situation; here empathy often becomes joined with sympathy. It is an example of the word's many meanings. If we, however, understand another person and utilize that understanding for our own benefit, this empathy may slip into exploitation rather than sympathy or caring. This sort of situation often leads people to abandon the word empathy and restrict it to a positive goal of activity. The neutral position of empathy as mere data gathering – i.e., understanding alone – is rarely adhered to.

Psychopathy: an indicator of knowing without feeling

Psychopaths or sociopaths are often quite good at reading other people. Such abilities may serve them well in manipulating others by convincing them that they are understood. Perhaps due to a particular form of brain organization, such individuals do not "feel" situations as others do, and this neurological deficiency reflects itself in a hypertrophied mind reading capacity along with a diminution in what may be thought of as "mature empathy" (Herve & Yuelle, 2007). Many therapists and analysts, in turn, feel unable to be empathic with this very general category of patients, although some can "read" them accurately but are simultaneously struck with a lack of feeling for them. It is important here to note that there are differing concepts of empathy.

This distinction serves as a valuable indicator of the need to clearly differentiate concepts such as a "theory of mind" (Premack & Woodruff, 1978), or "mentalization," (Fonagy & Target, 2006) from empathic inquiry. The emphasis on cognition may at times be so accurate that "knowing" becomes equated with "understanding." Heinz Kohut insisted that intuition was primarily a rapid form of cognition and was to be differentiated from empathy, which was an entrance into the inner lives of others (Kohut, 1971). In some patients the defense of intellectualization is more of an indicator of a particular inability to make emotional contact, much like a psychopath. However, in such patients, analytic work can enable them to better experience affect, while this may be a futile effort in psychopathy. Therefore we may introduce a significant parameter in our study of empathy with a question: is it alterable with treatment or is it fixed by certain neurological restraints? If it is the former then what and how shall further development be a part of an analytic or therapeutic pursuit?

Borderline personality disorders

It is probably not too far-fetched to conclude that, with a few exceptions, borderline personality disorders are spoken of in negative terms with descriptions that reflect difficulties in their treatment either by psychological or psychopharmacologic interventions. One rather universal appraisal of these patients underscores their poorly regulated emotions with outbursts of anger or other "affect storms" often explained by some form of "splitting" concept. Such patients are often very trying to analysts or therapists who may or may not be sensitive to these "affect storms." Indeed there seem to be analysts who are quite comfortable in the treatment of borderline personality disorders and therefore are positively characterized as being sensitive to the emotionality of such patients, while others feel quite unable and therefore unwilling to consider treating such patients. This emotional sensitivity may be called by some "empathy."

One may readily conclude that some analysts are able to be empathic with patients with borderline personality disorders while some feel quite at a loss in dealing with them. The major aspect of the problem, in contrast with that of psychopathic individuals, has to do with the intense and unpredictable and seemingly unrealistic emotional outbursts – outbursts that some analysts seem quite comfortable with and quite able to comprehend and interpret. This fact is perhaps indicative of the disparity that is revealed between differing empathic capacities that exist in certain patients as well as in certain analysts.

Autism

Although there is a good deal of controversy about the capacity of autistic individuals to be empathically accurate with others, there does seem to be some agreement that in severe autism there is a great inability to accurately recognize other people's thoughts and feelings (Baron-Cohen, 1995). This data seems to substantiate the likelihood that there are neurologic limitations to the ability to expand one's empathic ability, although there are no good data as to whether otherwise normal brains can develop increased empathic capacities. Perhaps autism can itself be categorized according to the ability to cognitively read others, as regards emotional comprehension, and further on to understand the psychological state of another – i.e., the combination of feeling and knowing. That is the neutral position of understanding and makes no additional motive of caring for or using one's empathy.

Projective identification and transference

When an appraisal of a particular psychological state is made by an analyst, it is usually assigned to a location by way of representing it. Thus, if a patient is angry because of feeling mistreated or misunderstood that state of affairs is explained by positioning it in the transference and so as being fundamentally derived from an earlier childhood experience. Of course if one feels that the patient is justified in such a set of feelings, it may be located as properly belonging to the ongoing relationship rather than it representing a misplacement. Without in any way diminishing the status of projective identification it *is* representative of a difference of opinion about location. If an analyst feels a particular set of feelings and understands them in one or another theoretical explanatory manner, it becomes necessary to determine if these feelings and meanings belong properly in the analyst or if an alternative site is more appropriate. Place has determinate role in empathic considerations.

To return to the angry patient an empathic appraisal might be:

1   You are angry because you are re-experiencing some mistreatment you felt you experienced as a child.
2   You are angry because I have done something to mistreat you.
3   You are making me angry to let me know how you feel.
4   You are unable to deal with your anger so you project it onto me.

No doubt there are other possible appraisals, and this should not be seen as an exercise in technique but rather as illustrative of how empathy requires emotion, cognition and location.

The psychoanalytic view of empathy

Any study of a line of development suggests a movement that is directed at an ideal goal that serves to reflect maturity. First presented by Ferenczi (1913) and later elaborated by Anna Freud (1965), the lines of development had direction that reflected health and maturity.

When modified by Gedo and Goldberg (1973) these lines were depicted as maturing both vertically and horizontally, and so the insistence on an ideal, normal or mature endpoint was necessarily reconsidered. However, the implicit and perhaps mistaken assumption in the presentation of these perspectives was one that assigned maturity to the analyst and the need for further development to the patient. The mythical ideal analyst would be empathic with all patients, and the mythical ideal analysis would result in similar competencies to the patient. Of course, this is not true. Not all analysts are empathic with all patients, and we often select an analyst who may work well with a certain class or category of patients and not with others. The more challenging question has to do with the goals or expectations for empathy from our patients.

To begin with considering the analyst, we recognize that among analysts there is a wide spectrum of empathic capacities, and we do attach a value judgment to these capacities. Our training programs do seem directed to the expansion of these abilities. However, there is often a somewhat fatalistic attitude as to the potential for growth. This is rarely seen as an area for further research and inquiry. Perhaps an expansion of the investigation on matches (Kantrowitz, 1996) would be able to better pinpoint the way we decide "who works well with what sort of patients" in the line of empathic development.

Impediments to empathic expansion

Given the admittedly valid biological constraints on one's empathic capacities, the usual explanations and reactions to a limited ability to be empathic are pejorative ones. Either one is too narcissistic to care about others or one was not brought up properly. The other explanation offered is often a moral one – i.e., they do not deserve understanding. One striking example of the latter is the general critical and unempathic attitude toward those Catholic priests who behaved badly. There is a sort of universal condemnation with rarely a modicum of effort to better understand them. More to the point is the sometimes explicit condemnation of the offender who is himself accused of lacking empathy toward those he has mistreated. Indeed, we are most unempathic with those who themselves suffer from this deficiency.

To move a step beyond the negative and critical stance taken toward the unempathic, there is a feeling that such persons can be helped to be more empathic with analysis or therapy. The antidote to the callous and cruel is available in the form of psychological intervention – for some but not all. And the dividing line is not very clear in terms of who can be helped. In terms of development, unempathic people are arrested at some ill-defined narcissistic stage and must be helped to re-activate a healthier pursuit involving caring about others. A working hypothesis for this ameliorative effort is that of removing the neurotic impediment and allowing development to proceed, or encouraging an identification with the empathic analyst, or perhaps uncovering a nascent and underlying empathy in order to reveal and express itself. Certainly not every patient becomes more empathic in the same manner. Some become more in touch with their feelings, some become more cognitively adept while some are better able to read their own and others' minds. Every analyst has probably had the experience of patients who were better able to feel things more clearly, to understand things better and to claim responsibility for their own thoughts. The movement from experiencing an empathic other – i.e., the analyst – to self-empathy is that of the path of an understanding other to self-understanding and is one that should ideally culminate in self-analysis.

And indifference

While there is often a positive correlation with the presence and use of empathy, there is regularly a negative one to indifference. Empathy is readily connected with "caring" and indifference with a form of callous "not caring." This negative appellation is somewhat tempered by the claim that indifference to events and situations that should evoke care or concern is a defense or protection against potentially upsetting feelings. We turn away from sad, painful or awful experiences because we cannot cope with the feelings. However, neither lack of feeling nor avoidance of feelings has much to say at all positively about indifference.

When someone is indifferent to a certain form of music or poetry, for example, we do not attach a negative connotation to that choice. In fact, it may become a decision of pride. There does seem to be a place for indifference that allows one to stay apart from concern without such a stance being seen as a failure of commitment. That distinction opens the door to the psychoanalytic preoccupation with neutrality, which appears to put a quantitative ranking on caring – i.e., we need to care, but not too much, and not too little. We must be empathic but must not over-identify or be dismissive. To merely state that one must be equidistant from id, ego and superego (Freud, 1946) may itself be indifferent to the problem of neutrality. Psychoanalysis must include both indifference and neutrality perhaps in a more careful inquiry of the development of empathy. They both are in need of greater attention.

Discussion

The popularity of empathy, as well as its almost universal acceptance as a virtue, has led to a looseness of usage and meaning. Psychoanalysis needs to establish its own comprehension of the term, and there may be no better way to accomplish this than to describe and formulate a developmental line for empathy. In a tradition laid down by Ferenczi and Anna Freud, our understanding of the mind is that of a product of change over time, and our understanding of the analytic method involves new ways of thinking about regression and progression in terms of nonlinear dynamic systems. With full recognition that we no longer think in terms of a single and progressive line (Coates, 1977), we do continue to think in terms of development. Earlier I wrote,

> The introduction of nonlinear, dynamic systems into our thinking about development allows even more for the recognition of distinct individual differences. A nonlinear system is one that has significant responses to minuscule changes, demonstrates emergent properties from elements that do not themselves contain that property, and in which the size of the input does not determine the size of the output. . . . We thereby rely more on descriptive efforts than on linear causal explanations. . . . With nonlinear systems, one can never determine beforehand the fate of an individual based on the usual considerations of development but must rather allow for the person to be a product that can only be explained and understood in retrospect.
> (Goldberg, 1999, p. 61)

With this in mind a developmental line for empathy may have a special relevance for psychoanalysis, since our capacities to be empathic with individual patients are necessarily variable, just as our patients' capacities for self-reflection and self-analysis are never of a single piece. The empathy of psychoanalysis is, of course, caused by and based upon "matched neural representations," but it goes beyond that to a discourse centered upon understanding the inner life of another, which involves knowing, feeling and proper positioning.

The equation of empathy with the phenomena of sharing feelings begins with the study of primate-sharing yawning (Cohen, 2010, p. 167) and is extended to the various discussions about a "theory of mind," which entails thinking about others. Empathy soon becomes a combination; i.e., perspective-taking or representing an idea about the internal life of someone, including oneself. Psychoanalysis adds the component of sustaining empathy over time (Goldberg, 2011b). Prolonged immersion in the life of another lends itself to the construction of a narrative that is organized according to the particulars of one or another analytic theory. Psychoanalysis can and should reclaim a special expertise about empathy by repositioning it in terms of its comprising a family of phenomena. In so doing we distinguish its role in the practice of psychoanalysis both in terms of its allowing a form of gathering data – i.e., learning about the inner lives of our patients – as well as in its therapeutic action – i.e., enabling people who feel understood to gain symptom relief by way of that connection. Whether or not that second step should remain a part of the definition of empathy remains problematic. We do characterize certain individuals as being empathic and using that empathy in a destructive manner.

We have learned that for many words and concepts there are no fixed or rigid definitions or meanings, but rather they can be understood only by way of their use. There is a book entitled *Wandering Significance* (Wilson, 2006), whose title nicely summarizes the journey of many of our ideas. There are multiple ways empathy is and should be understood and explained in terms of the particular community of users, and so the empathy of psychoanalysis has but a family resemblance to that describing the world of chimpanzees (Cohen, 2010). John Dewey and Ludwig Wittgenstein stressed that there is no more to the meaning of an expression than the overt use that we make of the expression (Quine, 1987, p. 130), and in psychoanalysis it is used to contain at least three features: emotion, cognition and perspective. These three components of empathy describe how we come to understand one another. As a shorthand we may often stress one or the other, but all are required to communicate the comprehension of others. Some feel it necessary to add a fourth component that goes beyond understanding with the activity of a positive purpose. We now embark upon the task of situating empathy in the practice of psychoanalysis.

Sustained empathy

Although the literature on empathy, no matter how one defines it, is enormous, often no effort is made to distinguish what one might call "ordinary empathy" or "common sense empathy" from "sustained empathy" (Aragno, 2008). This latter concept was one championed by Heinz Kohut in his explication of self psychology (Kohut, 1971), but it is regularly collapsed into some form of a general collection of ideas either about one's being empathic or else in teaching empathy or in having empathy for certain specific affect states. My goals in this chapter are to try to distinguish between these two conceptualizations of empathy – i.e., the ordinary, often instantaneous, and the sustained – and then to elaborate some distinctions peculiar to sustained empathy. The first task can be simplified by invoking a comparison between a single snapshot and a video; in other words, a time line is introduced to separate what seems to characterize most discussions and definitions of empathy in the ordinary sense from what distinguishes the lengthy immersion of one person in another's psychological state. The quality of this latter condition – what makes it unique, as well as the implications of the effort – is the aim of this chapter. Empathy over time is a qualitatively different phenomenon, more than the mere quantitative idea of temporal measurement would suggest. Understanding another person over time leads to a sequence of events, and that sequence for psychoanalysis offers a particular form of explanation.

Two meanings of sustained empathy

R.G. Collingwood is the philosopher of history who is credited with the insistence that all ideas, all facts, must be historicized. He felt that the past can be understood only by discovering the intentions of particular persons at particular times (Inglis, 2009). Thus one must put ideas in context as well as include the persons who capture these ideas, and the stories of those persons. Darwin, of course, taught us that all persons are results of a long series of historic events (Coyne, 2009). Sigmund Freud is the preeminent voice for stressing our need to see how someone got to where he or she is, yet we often read of being empathic with another person as if it were a slice of time rather than an unfolding story. This is especially true of much of the neurological study of empathy.

A recent article on electrical activity in the human brain demonstrated neuronal activation in Broca's area for speech production. It added the caveat:

> As is known for neurons in the visual cortex, the specific contribution of Broca's area may well vary with time, as a consequence of the different dynamic cortical networks in which it is embedded at different time slices. This fits well with the finding that Broca's area is not language specific, but is also recruited in the service of other cognitive domains, such as music and action and with the finding that its contribution to language processing crosses the boundaries of semantics, syntax, and phonology.
>
> (Hagoort & Levelt, 2009)

This echoes the contention that the data gathered by empathic connection, or indeed by any means, must be seen both in context and over time. Different time slices are but snapshots toward encompassing a complex set of meaning. The "here and now" is often a moment in time that may well be interesting but is best seen as a gateway portal to an arena of understanding. At times the "here and now" is accessed only by way of sustained empathy, but it regularly requires a place in an ongoing sequence of meanings. This is the first aspect of sustained empathy; i.e., it is a sequence of events and emotions and is never singular. The second aspect has to do with the impact of sustained empathy. A few case representations will follow. There is no doubt that the thesis may well appear simple and obvious to some. The point of the exercise is to underscore the fact that empathy usually but not always extends along time, while brain studies are more often than not representative of a moment in time.

Case report #1 – Charles: an example of fore-conception

Psychoanalysts approach case material differently than those who are not analytically oriented, and this approach is often described as one involving hermeneutics, or the science of interpretation. Its process is termed the hermeneutic circle. The process begins with the fact that we routinely know what we are looking for, or as Heidegger, the father of hermeneutics, would say, interpretation is grounded in a fore-conception (Heidegger, 1927, p. 141). The process that is suggested to achieve this grounding is a threefold one whose constituents are called (1) fore-having, (2) fore-sight and (3) fore-conception. This involves (1) a tentative knowing of what is to be uncovered or disclosed, followed by (2) an approach that makes things comprehensible and then (3) a grounding in a definite conception. These are the steps of the circle, and the circle is the structure of meaning (Goldberg, 2004, p. 205). This case is presented to illustrate how one listens both over time and with time in mind to gather meaning and so to participate in the hermeneutic circle.

Charles was a 52-year-old homosexual who lived alone and had no lasting significant partners. He came hesitantly into treatment because of a feeling that he was missing out on life. He was the fourth child of 10 children born to a Korean mother who spoke little English and a Caucasian father described as distant and aloof. After a series of weekly visits it was suggested that Charles begin an analysis, a suggestion to which he readily agreed. Initially the analysis went quite smoothly and Charles claimed to feel comfortable and hopeful. After one month Charles reported a dream that was interpreted to the mutual agreement of both Charles and his analyst. However, the day following this interpretation Charles insisted on sitting up. He challenged his analyst with being up to something insidious and announced that he was quitting his analysis. Inexplicably, after a short absence, Charles returned to analysis and resumed with the calmer conduct of an analysand who was content with the process and the results. Along with this claim of contentment and improvement, after another presentation of a dream and its successful interpretation, there occurred a similar set of behavior with accusations and separation. This sequence of calm followed by a severe change of emotional state with subsequent recovery only to repeat itself presented a befuddlement to the analyst, who could see general and gradual improvement only to be met by a seeming undoing of all that had been accomplished.

When this case was presented to a conference to discuss how the issue of its analyzability was so mistaken and just why the treatment was so punctuated with both hope and despair someone suggested an explanation. This consisted of an effort to be empathic across time. The hypothesis that was offered could certainly not be confirmed but consisted of the possibility that the patient was demonstrating what it was like to be one of 10 children and to have his closeness to his mother repeatedly interrupted by the birth of another child. No sooner was he content than his life was turned upside down. Thus one listener chose a developmental conception to guide her listening.

This case is not presented as a necessarily accurate portrayal of this patient but rather as a reminder that we do regularly employ empathy along a time line, and we think of cases in a narrative fashion. When the patient sat up and behaved in what the analyst termed a "paranoid" manner, one could be empathic with his rage, and one could conceptualize it as a changed emotional state following an interpretation. However, being empathic in the "here and now" is but an entrance into the overall task of understanding. The hermeneutic circle, which is often referred to as representing the activity of psychoanalysis (Chapter 2), is seen as a back-and-forth process that is modified by each of the participants. The ability to sustain one's empathy during the course of the analytic process allows the analyst to replicate the history of an individual's developmental life. Of course this is not definitive as such but it serves to distinguish psychoanalysis as an interpretive science that employs sustained empathy.

Case report #2 – Elizabeth: the modification of the hermeneutic circle as new material is disclosed

After her analyst moved to another city Elizabeth was urged to resume analysis without delay, and she did so without a moment's hesitation. She presented herself to her new analyst as a grief stricken patient who had lost someone of great importance to her and whose loss seemed to generate an intense anger that surprisingly soon became directed at her new analyst. This replacement analyst felt that he could be empathic with this new patient's rage and bereavement, but Elizabeth would have none of this "empathy," which she scoffed at and ridiculed. Indeed she soon revealed that she was not so much angry at the supposed unfortunate demise of her old analyst, but rather she had constructed a complex justification for her rage at her new and seemingly innocent present analyst.

Elizabeth's new analyst readily moved from his empathic stance of an attempt at understanding the anger of his new patient to one of puzzlement along with his own anger. Once again we see that empathy is never a matching of feelings and the hoped-for understanding that results, but rather is a complex configuration that is a story told over time.

The story that did explain Elizabeth's rage was revealed after some period of treatment and both surprised and dismayed her new analyst. Elizabeth felt that her old analyst was not helping her at all, and yet she was completely unable to extricate herself from him. She had constructed an imaginary scenario that consisted of her new analyst or someone like him interfering with the ongoing process of her old analysis and either setting the old analyst straight or else rescuing her from this unfortunate entanglement. She was thus justifiably (to her) angry at her new analyst's failure to rescue her. Of course this particular fantasy could not possibly have a basis in fact nor could it have been readily accessed by an immediate empathic stance.

Empathy may at times offer an entry point into the hermeneutic circle, but it may also serve as a totally erroneous door to nowhere. Empathy is usually layered as well as sequential. Beneath Elizabeth's anger was her severe and painful disappointment at the failure of her analyst and also at her parents. There is no way one could gain access to this experience save by the analytic process, and this necessarily takes place over time. Thus her analyst had to, in turn, be empathic with her rage, her life experiences leading to her fantasy and her subsequent disappointment. Thus one needs to sustain and modify empathy to achieve understanding. Empathy changes over time and alone it explains little. Empathy is the data that must be carefully organized in terms of cause and effect, sequencing and goals (Goldberg, 2011b, p. 130). If seen in isolation it is but a form of attunement or affective resonance that need have no meaning on its own. Its universality should not be taken to be indicative of any particular therapeutic benefit, inasmuch as here too one must not see it in isolation. Empathy can be good, bad or indifferent, and it is to this we now turn, once again keeping in mind the need for it to be lasting and often beneficial.

Case report #3 – Mike: the time line in developing the hermeneutic circle

Mike came into treatment in the midst of a contentious divorce, which became resolved only after months of acrimony. The marriage might properly be characterized by Mike's taking care of everything up and to including making dinner when he returned from work, while his wife merely did as little as possible. Of course this was Mike's own characterization of his marriage, but I doubt that it was very far from the truth. I became more convinced of this state of affairs when Mike told of incident after incident of his competence. He carefully explained to the car mechanic exactly what was wrong with a malfunctioning foreign car. He was right, and the mechanic was impressed and grateful. He patiently explained to a coworker how to organize a particularly complex set of ideas and, once again, he showed his extraordinary ability to solve complex problems. In all of these examples, and there were many, Mike was exceedingly polite and gracious, although in his recounting of these stories of his super competence Mike was less than charitable toward his listeners, who ranged from positions best described as pupils or colleagues or adversaries. Mostly as a whole they were tolerated.

It was not difficult to empathize with Mike's feeling of superiority over his incompetent fellows. He was sometimes exasperated, sometimes pleased, sometimes enraged and sometimes even surprised at his own cleverness. Mike could discuss just how he affected others, including his now ex-wife, along with the new women that he began courting. These discussions led to Mike's recounting of a childhood that seemed to be representative of two dominant traits. One was that of learning from an uncle how to fix all sorts of complex machines. The other was of being absolutely terrible at sports, from striking out at baseball game after game to never really learning how to swim. One strand was of competence and the other of failure. Beneath Mike's never-ending showing others how to do things was the little boy who could hardly do anything very well. Getting in touch with the helpless child whom Mike continually wished to disown was an empathic stance that was achieved over time. It lay under, or hidden from, the extremely competent person who could more immediately be apprehended and recognized. When Mike himself was able to be in touch with his own fear of his own incompetence, he became almost apologetic about knowing so much.

The therapeutic effect of sustained empathy

We have seen sustained empathy as aiming not for a moment of meaning but as an extended explanation of sometimes contradictory and hidden meanings.

The therapeutic effect of this sustained empathy has to do with the impact of empathy on the one who is the target of the empathy, sometimes called the empathasand. There is much evidence in day-to-day practice that people feel better when they feel understood, and there are a multitude of explanations for this positive state. Different psychoanalytic theories offer different explanations ranging from the release of the repressed, ascription to classical theory, to a host of other such possibilities. However, there does seem to be a difference between the pleasure one receives when, say, a dream is interpreted and so seems to make sense, and that of the ongoing contentment and satisfaction experienced when one feels connected and understood over time. Agosta makes the claim that one gains one's own feeling of humanness from another human being, and so he declares that "empathy is the foundation of human community where 'community' means 'being with one another in human interrelation'" (2009, p. xiv).

This connectedness of one person to one or more others is to be considered in both its short-term and long-term effects. They seem both similar and different. Short-term effects are commonly thought of as a cognitive achievement, such as when one gains insight following interpretation. Long-term effects need not have a significant cognitive dimension but may occur when one feels a connection to a person or a group that is seen as sustaining or fulfilling. In truth, a multitude of studies outside of psychoanalysis have been conducted to demonstrate the positive therapeutic effects that result from this feeling of belonging or participating.

The nonanalytic studies that have concerned themselves with the benefits of human relatedness range from those that measure the biological mechanisms that explain the positive association between social integration and physical health (Hawkley & Cacioppo, 2003) to the studies of particular diseases and social relationships (Bae et al., 2001). They also extend to the hormonal biochemical issues involved in the connecting of one person to others (Van Anders & Watson, 2007). There is quite an accumulation of research and data on the full range of the issues, involving the positive as well as the negative effects of social relationships (Cacioppo et al., 2002). What is lacking in all of the data is an adequate psychoanalytic or even a psychological explanation of the significant correlation between emotional well-being and social networks. Every level but the unconscious is mentioned and measured, but sustained empathy and its psychological impact is bypassed. Basch felt it necessary to use the term "empathic understanding" to emphasize the psychological, but our use of the term "sustained empathy" is really shorthand for sustained empathic understanding.

Since sustained empathy is a cornerstone of psychoanalytic self psychology, it is no surprise that one of the best, but not the only, explanation for its therapeutic effect comes from that theoretical vision. The theory that is proposed is one built around the selfobject or another person serving as a fundamental part of the self. Thus the person utilizes others as psychic structure. The person is constituted and sustained by these relationships to others; therefore, self psychology is essentially a one person psychology that explains how others become aspects of one's self. Deficiencies in one's self become filled by others, and so a selfobject relationship leads to a feeling of self-integration. One is made whole by others.

If one considers a social relationship such as marriage and evaluates the positive and negative effects of such a connection, the results can be seen or explained at many levels. For instance, married men are more likely to adhere with recommendations for screening colonoscopy than unmarried men (Denberg et al., 2005). If one studies this level of interaction at the most obvious level then one may miss the content of the psychological level. One person might say that his or her spouse will be pleased or proud and may mirror the activity of such adherence. Such connections may be enduring and maintain those personal psychological experiences that are necessary for self-esteem regulation. Another way of describing these connections is by seeing them as sustaining empathic relationships. Such enduring relationships serve to maintain an integrated self.

Just as the one who is the target of sustained empathy is to be evaluated as to its ameliorative or negative aspects, so too must we consider the emotional impact on the so-called empathizer. It may be no easy task to persist in an empathic connection. Some time ago I attended a case conference at which a severe schizophrenic patient was discussed in detail. One of the listeners spoke up to say that the material was simply too painful to stay with and so thereby illustrated the terrible state in which the patient lived. Empathy is always a two-way venture and may exert either a cost or a benefit to each participant. One is being asked to function as a particular selfobject when one enters into an empathic connection. Often this match is fortuitous, as when a patient asks to be mirrored and an analyst recognizes and responds as such. Sometimes the match is unworkable, as when a patient wishes to be mirrored and an analyst needs to be idealized. More important than the simple fitting together of a desired selfobject is the need to persist in the linkage over time. Once again we must distinguish between a temporal slice of understanding and an enduring experience of sustained empathy.

Prerequisites for sustained empathy

Thomas Metzinger is a philosopher and scientist who suggests that we extend the concept of empathy to account for all of the different aspects of expressive behavior enabling us to establish a meaningful link with others. He proposes the term "shared manifold" to capture the phenomenological, functional and subpersonal levels of human connection. The phenomenological is the conscious sense of similarity, the functional is the actions or emotions we observe in others and the subpersonal is the activity of mirroring neural circuits (Metzinger, 2009, p. 175). Similar to most of those who liken empathy to something akin to textual reading, Metzinger assumes that we all read the same sentence and possibly interpret it differently. Similarly, in a recent column in a newspaper, the columnist listed a series of connections between the brain and psychology ranging from the situation of "menace" activating the amygdala to the anterior cingulated mediating pain, and thereupon made a leap to how such studies may someday tell us "how people really are" (Brooks, 2009). These snapshots of human interaction construct a scenario in which the observer is outside of the interaction. They fail to see that all empathy, both the immediate and the sustained, is a two-way street.

When we do observe as part of the interaction, we realize that a certain set of demands are placed upon the person practicing sustained empathy. The first of these demands is the requirement to avoid premature closure. This particular delay in decision-making may require us to tolerate anxiety or any positive or negative affect associated with the particular memories and feelings aroused in us. Thus, for instance, the connection of so-called menacing with the amygdala might ordinarily stimulate our own functional aspect that Metzinger categorizes and so may lead us in a direction that is not correlative to that of the person with whom we wish to be empathic. Indeed, there is no way that anyone can or should attempt to maintain a completely neutral position in sustaining one's empathy. Each category of Metzinger's should be calibrated in terms of the input of the empathizer. Thus the second requirement for sustained empathy is to understand and manage the fantasies that are stimulated in the person who is being empathic. These fantasies of course are valuable contributions to be considered in the practice of empathy and are not to be dismissed or condemned but rather to assist in understanding. If one is able to resist premature closure and allow a recognition of one's personal contribution in this effort to understand another, then a third demand or requirement comes to the fore. This last needed activity is that of establishing one's own time line and so breaking the empathic connection.

This act of empathic disruption is said to be significant in terms of psychological growth and/or possible insight. Much of that growth is predicated on the ability to re-activate an empathic connection that enables one to evaluate, again over time, the impact of sustained empathy, followed by empathic disruption and its subsequent reconnection. Sustaining empathy is an act of careful and deliberate effort that imputes a responsibility to one that goes far beyond Metzinger's threefold levels. All observation imputes participation, and this participation varies from one observer to another.

And mentalization, etc.

As noted earlier, mentalization is a term introduced to offer a theory of how mind reading is developed. It is said to be dependent on secure attachment (Fonagy et al., 2002); types of attachment/attachment patterns in turn are said to be correlated with certain forms of pathology. Empathy is considered by some to be a form of mentalization (Frith & Wolpert, 2004, p. 115) and so joins a host of other words or phrases employed to make better sense of how people communicate with and understand others. At times mentalization is restricted to observed behavior outside of language (Frith & Wolpert, 2004, p. 48), and at times an effort is made to explain it in a neuroscientific manner.

Mentalization is but one of a number of words and phrases that aim to differentiate a particular activity involved in a therapeutic relationship. Just as Basch preferred empathic understanding, others prefer empathic immersion, attunement or stance, or similar variations on this single theme. The only distinction attributed to sustained empathy is that of the time line that is required for determining the meaning that is formulated and the interpretation that is offered.

One can readily see how different psychoanalytic perspectives, each with its own terms and concepts, from interpersonal to intersubjective to relational, derive from the question: how does one person manage to determine what is going on in another person, and what is the impact of any and all such connections? It is probably futile to attempt to differentiate these terms one from another, and it is likely that sustained empathy is a factor in them all. However, inasmuch as common or ordinary empathy is part and parcel of a host of psychological operations as well, it is important to carve out a particular activity that is the domain of sustained empathy. It is not mere listening. It is not mere mind reading. It is not merely having a relationship either interpersonal or intersubjective. It is all of these that lead to understanding over time, and time is the crucial ingredient.

Discussion

If we see sustained empathy as an ongoing fitting together of selfobjects to aid in self-integration, we can extrapolate this form of connection from its role in psychoanalysis to all forms of social interactions. The selfobject transferences that arise in psychoanalysis result from a sustained empathic stance similar in type and form to those that emerge in all sorts of meaningful social relationships. Such enduring linkages differ in one crucial way in terms of effective treatment and something like mere group membership.

This difference is underlined by the analytic process, which aims to enable the person not to be independent of others but rather to form stable and sustainable empathic connections outside of the analytic situation. The analytic experience is not one to be copied or imitated outside of the consulting room but rather to be seen as enabling one to form sustaining empathic connections. It would be a fundamental error to see the loneliness of a person as treated by the companionship of a therapist or analyst, since loneliness is not a product of the lack of companionship as much as the inability to attain and retain companionship. To study isolation as an unfortunate situation, which it surely is, does not allow one to see how sustained empathy is an achievement and not a happenstance.

Summary and conclusions

Empathy can be studied as a product of certain hormonal changes, as a particular form of brain activity, as a vital ingredient in the formation of social relationships and even in its presence in animals from whales to non-human primates (de Waal, 2009). Inasmuch as these are different levels of inquiry, there is a danger that the term may be either trivialized to the point of losing its meaning or glorified to a point of panacea for all sorts of problems.

In a somewhat abbreviated form, empathy is a method of data gathering – i.e., a category of mind reading. When we do read what we believe is going on in another person's mind the data that we gather is regularly treated not so much as a bunch of words or sentences but rather as ideas laden with meaning. Empathy is thus not a mere registration of thoughts and feelings but is a complex configuration best thought of as a story or narrative. If we add the historical component to what we read in another's mind we pursue empathy along a time line and thus move to sustained empathy. This activity of gathering information about another often seems, in itself, to change the nature of the information that we gather. Not only does the observer affect the data gathered, but the very act of observation changes both the observer along with the observed. Thus we conclude that some empathy is thought of as a single slice of information that is layered and some as a lengthy set of information with a cause, a sequence and a goal. It is suggested that the latter be separately categorized as sustained empathy and studied separately for its own therapeutic effects.

If one studies empathy primarily on the level of psychology, psychoanalytic self psychology offers a particular form of insight as to why it has both salutary and negative effects. The empathizer, or the one offering empathy, is experienced as a necessary selfobject by the empathasand, or the one receiving the empathy. Over time this meeting and matching of self and selfobject aids in self-integration and self-esteem regulation and so leads to a feeling of well-being. This feeling may, on occasion, be shared by both participants. Thus people feel better when understood and achieve an added feeling of self-cohesion when understood over time. Sustained empathy has a status that is qualitatively different than a short-term empathic connection with another person. Indeed, it may be one of the defining characteristics of human beings.

Chapter 8

Self-empathy

The idea of possessing a set of feelings and thoughts about oneself is not an unusual one inasmuch as we may feel sorry, upset or even pleased with ourselves in normal discourse. Yet the idea of an activity designed for understanding ourselves is a bit uncommon save for an enterprise such as psychoanalysis, which seems to be primarily directed at self-understanding, albeit in a special way. We are said to be self-interpreting animals (Taylor, 2002) since there is little doubt that we regularly explain and interpret things, events and people to ourselves. Any patient with gastric reflux problems interprets tonight's heartburn as resulting from this evening's pizza, just as any sport enthusiast interprets Monday's discontent as due to the favorite team's loss on Sunday. Humans may be unique amongst animals in this form of relentless self-inquiry. Of course all animals anticipate and plan how to behave in certain situations, but it is difficult to determine with certainty if a predator such as a lion contemplates the details of his or her last kill. However, few of us fail to review and/or regret our latest stupid or brilliant statement.

Extrospection is the opposite of introspection and is the activity of examining what is situated outside of us. Although we may primarily inspect things in order to gather information, we also regularly need to determine the meanings of this information, and so our extrospection is a collection of facts and meanings. Introspection, however, is devoted to collecting facts and meanings about our inner life. It does seem to be the case that the facts and meanings gathered from extrospection often become gathered into a different set of facts and truths – e.g. if the additive in that new gasoline gives more miles to the gallon then it will cost less money to drive from point A to point B – while the facts and meanings from introspection seem to always call for further interpretation – e.g. how do I feel about going to point B? One may disagree about this distinction, but for the most part empathic inquiry is an act of introspection and is primarily directed at understanding another by way of vicarious introspection *or* at understanding one's self by way of self-empathy.

For the most part our thinking about and interpreting our inner states is a continual activity akin to a series of snapshots with only a rare sustained inquiry over time. The introduction of a time line to self-scrutiny is one parameter that alters our self-interpreting activity while the added component of unconscious factors makes for a radical alteration in our introspection. Self-empathy requires thinking about one's self over time with the contribution of our awareness of the participation of the salient unconscious factors. No doubt some aspects of self-empathy operate to some degree in almost everyone, but it becomes markedly modified with the introduction of a time line plus those contributions that are not readily available to conscious contemplation – i.e., the unconscious.

The time line

H. had read in a college magazine that a small church on the campus was being demolished to make way for a new administration building. H. felt sad at reading of this alteration of the college that she had known, and she likewise remembered that she had attended a sermon given by Martin Luther King in that very church. Her sadness became mixed with anger as she contemplated the loss and replacement of this revered building. She thought about her feelings, wondered about their justification and was able to momentarily step to the side to consider herself along this time line. In fact, most people do something like this quite often, with perhaps a few insisting on concerning themselves only with the "here and now." It is probably almost impossible to eliminate our routine historicizing of events, and the act of examining and understanding our selves more often than not changes the mere snapshot to an extended video presentation – i.e., a narrative.

Unconscious factors

J. could not understand why anyone would care or be upset about the destruction of a church for a new administrative building. He said that he might be able to sympathize with H. over the prospect of the loss of that building with its storehouse of memories, but he could not feel much about it, since new things always replaced older ones. It was not that J. was unacquainted with loss, but rather that it was merely a fact that had to be accepted. J. was an only child, and in his personal analysis it was suggested to him that perhaps replacement and displacement meant more to H. than it did to J. J. could readily see how H., who had both older and younger siblings, might feel differently about the very idea of replacement, but he simply could not find that feeling in himself. He could cognitively understand H., but felt that he could not empathize with her. J. spoke with L., who was the youngest in his family, and L. said that he was possessed of a totally different set of feelings about displacement since he could empathize with his older siblings whom he had displaced. Thus J. was confronted with an unwelcome fact that seemed to limit one's empathy to a personal storehouse of experience and feelings that we have endured and so prevented us from completely understanding others. Surely it was easier to identify with the known, but did one have to suffer in order to help others who are suffering? However, J. was asked to consider the possibility that he might indeed know what he, too, easily claimed not to know. He remembered that time he had lost the first stand position in the band in which he played. He had only the briefest moment of anger until he realized that this new player was simply better at playing his instrument. J. wondered if anyone could go through life without somehow, somewhere having the feelings that we all seem to have – sometimes only slightly and briefly and sometimes for too long a time and too intensively. At this point J. realized that at times he simply did not want to be empathic, to understand another and so to understand himself. He did not like himself as the person who was hurt and angry at his musical demotion. He preferred the J. who was able to take things as they come without rancor and disappointment, and he surely could not be both angry and reasonable at the same time. He bet H. could also be quite reasonable about the need for a new administration building. The only conclusion that he could reach was that we were all a bundle of contradictions in search of plain, unalterable truths. J. began to include features for interpretation that were not immediately and readily available. Once we have access to our psychology beyond the superficial, we embark on an introspection of preconscious and unconscious factors that alter our interpretations in the form of a depth psychology.

We are confronted with the truth about meaning; i.e., it is multiple and often lacking in clarity. One analysand said that his analysis led him to the difficult position of always seeing things from his opponent's point of view. He said that his expanded empathy resulted in a certain form of paralysis as understanding prevented divisiveness. This form of introspection seemed to have its own problems.

The limitations of self-empathy

We all clearly need to deal with the inherent limitations directed at understanding ourselves and others, and these limitations usually derive from a particular moral posture that we assume. We may well be able to understand the motives and inclinations of a pedophile and even share a similar but slight predilection in ourselves. Yet we more strongly feel it to be wrong, and we may even justify our moral stance by being empathic with the victim. It is important that our self-empathy settle somewhere between absolute positions of right and wrong and relative positions of indecision. This decision about the determination of a stable position for ourselves is also applicable to the particular school of psychoanalytic thinking that we choose to join. We may well understand other ways of thinking, but we need a moral compass to guide us to what is assumed to be the correct way, all the while recognizing that the word "correct" merely means the most useful. One can conclude from this that there is no way to be empathic with another without a matching effort of self-empathy. If a patient has a strong feeling about a significant person in his/her life and after some analysis better understands that person and so modifies his/her original strong position, then that patient necessarily feels differently about him/herself. Indeed there is no way to understand another without a change in one's self and a corresponding evaluation of the degree of change allowable for oneself. We need to examine those elements of self-regulation and self-stability that allow for understanding without going overboard in agreement even to the point of conversion.

It is at this point of what we might call self-maintenance that we see the significance of our being constituted by our selfobjects. They serve to sustain us by a continual feedback of mirroring and idealization. That is why a community of those who share beliefs is a necessary underpinning to allow an empathic understanding of ourselves and others. It should come as no surprise that young soldiers without a firm sense of self are easily indoctrinated into the killing of others without a clear sense of the moral constraints concomitant with the very idea of murder. They cannot be empathic with the enemy unless and until they are able to understand themselves, and this is often reduced to the mere act of killing, reinforced by belonging to a community of killers. We cannot expect them to know whom to kill and whom not to kill (including themselves) unless and until they themselves know who they are.

The need for self-empathy

It may or may not seem obvious that the operation of self-empathy is a continual operant need in our daily lives as well as in that of psychoanalytic inquiry. One way to characterize psychoanalysis is that of the analyst involved in understanding and articulating what is understood to the patient. The latter, of course, may understand himself or herself to some degree, and the analyst strives to join and then go beyond this somewhat limited understanding. We have earlier noted how this feeling of being understood may be therapeutic, but the added contribution of seeing what had previously been unseen allows for the transfer and incorporation of the analyst's empathy to the patient and the ensuing realization of self-empathy in the patient.

Case reports

These are amalgam cases that are not meant to be representative of correct technique but rather as illustrative of the point of the need for self-empathy.

Lisa was an unmarried woman who enjoyed a high academic standing in a very specialized field and who reported a series of unsatisfactory attempts to establish lasting relationships. In a nutshell Lisa felt that she was misunderstood no matter how hard she struggled to have things otherwise. She found herself involved with men who expected things of her that she could not deliver and who in turn continually fell short of Lisa's own expectations. Lisa eagerly began an analysis and was an ideally cooperative patient who over some time periodically befuddled her analyst. No matter how hard her analyst tried to find some sort of psychoanalytic theory to explain Lisa, he found himself at a continual loss in his effort to construct a formulation that made sense. At long last the analyst realized that Lisa and he were recreating a childhood with a mother and father who could not comprehend how to be parents and certainly not how to parent Lisa. She was thus launched on a life of not being understood. In the transference, Lisa was joined by an eager analyst who could entertain only various irrelevant explanations of his patient. When this state of befuddlement was interpreted to and for Lisa, she felt a flood of relief when her ever-present status of being misunderstood was finally understood as a recreation of her childhood dilemma.

Of course the analysis of Lisa required much more than the simple idea, but it did serve to highlight the ever-present feeling of isolation and loneliness of Lisa along with the pressing need of the analyst to construct a workable explanation. Lisa was also resistant to the idea of being understood, since in a rather perverse sense she expected to be misunderstood and had organized her life around this fundamental concept. In her analysis Lisa moved from the feelings of hurt and anger at ever being misunderstood to the recognition of what those feelings essentially meant to her. The understanding of the need to repeat the feeling of being misunderstood resulted in the search for selfobjects who understood her, a rather striking achievement for Lisa, and an effort that had previously come to naught.

When Lisa was involved in a relationship with a man, she found herself making every effort to understand her partner, who, in turn, always seemed to get Lisa wrong. Of course Lisa represents an extreme of a process that is familiar to all analysts who live through the cycle of misunderstanding or befuddlement, followed by understanding with great relief, only to usher in a new phase of not understanding. Ultimately patients take over the cycle as they proceed to master the process by assuming the self-analytic function.

Every now and then we encounter a patient who seems unable to practice self-analysis. Rosa was such a patient who, while also eager and compliant, seemed either reluctant or ill-equipped to "think for herself." She would wait patiently for "answers," as if primed for someone else to do her thinking for her. Rosa assumed the posture of a perpetual student who admired her teachers but who never hoped to emulate them. In her family there were a number of rules for living, and Rosa told of literally being raised by a rule book. Her parents were not particularly authoritarian, but they were bereft of explanation – i.e., things were done because that was what was done. There were rules for everything and little space for inquiry or wonder. Rosa did not seem particularly interested in causal explanations, but she did demand answers. Especially in the activity of self-analysis and certainly in that of self-empathy was Rosa lacking interest. Curiosity was confined to learning the rule or procedure for performance, but not to the investigation of causes or reasons. Rosa smiled at talk of introspection, inasmuch as it seemed to her to be a delaying tactic on the road to the correct answer. Analysis was not for her.

Perhaps most children grow up in a mixture of rules and explanations as well as in one of understanding and misunderstanding. So, too, do many patients struggle with wanting advice on living versus wanting to understand themselves and others. Analysts too struggle with the urge to give advice versus that of achieving understanding. One test of analyzability surely is directed at a curiosity of why we do and feel what we do, and one task that becomes paramount in some analyses is that of development and fostering the wish to know our inner selves.

Self-empathy as an antidote to helplessness

The self-analytic function is usually thought of as an activity designed to assist an individual who is upset or in conflict over a particular issue or at a particular moment in time in order to make conscious whatever unconscious fantasy or memory is responsible for the discomfort. It is also usually thought of as cognitive, and as a more or less emerging or time-limited effort. Self-empathy can be considered as more of a continual self-assessment that ideally allows for comfort in a variety of circumstances.

Any symptom from a headache to a toothache to anxiety is apt to put one in the temporary position of a victim who is experiencing an unwelcome discomfort. The victim naturally moves to master the discomfort by doing something ranging from taking an aspirin or getting a tooth pulled or striving to analyze the source of the anxiety; i.e., one moves from a helpless posture to one of mastery. Even the beginning act of investigating and planning to relieve the symptom moves one from a passive position of enduring the discomfort or suffering to an active one of taking charge. As one understands the origin or source or any factor that initiates the symptom, one moves toward controlling and managing the discomfort with a partial diminution of the discomfort. Once conquered, we may be at ease. For psychological issues this feeling of contentment should involve an associated feeling of self-understanding, often characterized as being comfortable in one's own skin. That is self-empathy as a cure for feeling lost and at the mercy of external forces – i.e., helpless and impotent.

Discussion

The sequence of understanding on both a cognitive and affective level when applied to another person's inner life is defined as empathy. Psychoanalysis enables some persons to be more empathic with others, just as the myriad experiences of living do the same. Our day-to-day understanding of others also applies to ourselves, and this function is that of self-empathy. Our feelings and appraisals of ourselves reflect the gamut of feelings and appraisals of others and often include a further step of considering the meanings and implications of these conclusions. If we are able to incorporate the unconscious contributions of these ingredients we can be more empathic with others, and the same additive factors apply to ourselves. An analyst involved with Lisa and Rosa might well be irritated at not understanding the one and not enlisting the other in his project. This analyst would have to understand the sources of her irritation and frustration to not only be empathic with Lisa and Rosa but to do the same for herself. Her personal analysis and the continual self-analysis allow her to maintain and sustain the self-empathy that is required.

Summary

Self-empathy may be thought of as an enduring end stage of relative comfort and equanimity that results from understanding oneself over time. Although it is not a single moment of achievement, and does require a maintenance program of self-analysis, it results in an amalgam of ethical and moral positions, along with a recognition of personal ambitions and limitations. One must be able to live with and reckon with a life of both discontent and satisfaction, and self-empathy should never be thought of as a lasting state of self-satisfaction. Therefore any persistent state of dislike or dissatisfaction is a call for a better understanding of one's self. Indeed we are launched on a voyage of wonderment with psychoanalysis serving as an added tool in the adventure. Yet our understanding of ourselves may not extend itself to an understanding of others.

Part III

Clinical examples of the special role of psychoanalysis

Introduction to Part III

One cannot make a clear and coherent claim for the difference and distinction in psychoanalytic inquiry without first recognizing the present state of disarray in the many voices of psychoanalysis, voices that are collected in the allowable and recognizable state of "pluralism." The following section is aimed at presenting a unique view that psychoanalysis has on psychological differences and so of psychopathology, all the while noting that the different "schools" of analysis differ in their way of evaluating and discussing psychopathology.

When Bergmann openly discussed the danger of psychoanalysis becoming a "latter-day Tower of Babel" (Bergmann, 1993, p. 929), he urged that a new generation of analysts be educated to see psychoanalysis from a historical viewpoint. Since that time we can see only further sets of differences and more and more lack of communication between these competing "voices" in analysis.

The following section is an effort to both recognize the differences and seek a commonality in the understanding of mental disorders.

The field of psychoanalysis may be said to hardly be a proper home for mental illness inasmuch as it seems filled with its own form of instability. Psychoanalysis is composed of competing and squabbling schools with endless arguments about who practices authentic analysis, who is most faithful to Freud and, perhaps most significantly, who really helps those in need. Relying as we do on evidence-based empirical studies there is little hope for the emergence of a champion as the best school or theory of psychoanalysis. More importantly, the close affiliation with psychiatry has made a requirement for a champion to be that of effectiveness in therapy, since there is no other yardstick available to evaluate our competing schools. When Holzman (1985) made his case for our own consideration of the science of psychoanalysis as being more than independent of psychiatry, he unfortunately used the standards of empirical science to track hermeneutics, and he therefore insisted upon "validation" for psychoanalysis. However, it is unlikely that interpretations and meanings will ever fulfill that demand. Differing interpretations of the Bible are never "validated" as much as accepted on the basis of statements such as "sounds right" or "makes the best sense to me." As much as one may yearn for "proof" of "what it really means," we are ever having to reckon with rather vague concepts such as "opinion" or "personal belief."

If we carry over those vague concepts to the competing schools of psychoanalysis we will be forced to abandon a hope for a truly universal or general theory that would qualify for an ecumenical set of ideas. Rather we might settle for tolerance of differences with the probable shared belief that each of us feels we know best. Of course, over time one set of beliefs may dominate, but this result can hardly claim to be proof of truth and more likely to be the result of political forces. However, there must surely be some commonality to these various "schools," even though it need not be a common technique or vocabulary.

After a rapid review of the aforementioned reason to abandon a hoped-for single agreement on the meaning of a psychological state, some examples of a psychoanalytic approach will be offered.

Imagine the following scenario: a middle-aged man, who at one time had seen a psychoanalyst for therapy in the midst of an unhappy marriage, calls for an appointment. He wishes to discuss his feelings about the impending marriage of a daughter who was born during that unhappy marriage and whom he has not seen or heard from for many years. He has a brief but cordial meeting with this estranged daughter and her prospective husband prior to the wedding to which he is invited. He talks to his analyst of this meeting and his feelings about this daughter's mother as he prepares to attend the wedding. He decides to gift the daughter with a check large enough to cover the expenses of the wedding, which he attends with his new wife. He reports back to his analyst that he was intensely disappointed with the reaction of his daughter at the wedding inasmuch as he felt that he was a mere "prop." She hardly reacted to his presence. He tells the analyst that he has decided to cancel the check that he had sent.

Now despite what the reader may think of this man and his behavior, it is in no way able to be characterized in any psychiatric category listed in DSM-5. Yet it is surely the stuff of all psychotherapists' lives. While this man who is now a patient may feel angry at his daughter or perhaps ashamed at his own behavior or remorse at this failure as a father, there is surely no medication to relieve him or perhaps even no advice to give him. Rather, he needs to discuss and be aware of what this long absent daughter, her wedding and his appearance meant to him. This is the province of psychoanalytically oriented psychotherapy and well illustrates how psychiatry and psychoanalysis often have little in common. The question "what does this mean to you?" differs markedly from a concern with a well-functioning or damaged brain. This man employed his brain in a particular form of behavior that warrants study quite aside from any pathology that may be attributed to the brain. This behavior had to do with his self-esteem and a complex investigation of what his daughter and his being her father meant to him. Rather than facts to be examined, the larger area of meaning and the agency of self need to be explained.

What follows the interlude about positioning psychoanalysis to be a proper manager of mental life are two examples of inquiry that essentially have little to do with psychiatry and much to do with psychoanalysis. People who are enthusiastic or angry need not be, indeed often are not, mentally ill. The concept of being "sick" has often been distorted to mean deviations from some norm that is often based upon a moral or social standard that is constructed upon a personal basis. We operate best in a position of empathic understanding free of the bias of illness.

Chapter 9

The danger in diversity

Introduction

An announcement that visual imagery during sleep could now be objectively decided and so read out from visual cortical activity patterns (Horikawa et al., 2013) makes possible the discovery of the linkage between verbal reports of dreams and the brain activity associated with these reports. The study can be likened to Freud's distinction between a manifest dream and the latent meaning or content of a dream (Freud, 1900). It again highlights the distinction between the empirical findings of a dream – i.e., what is represented – and the interpretation of a dream – i.e., what it means. Of course one may use a dream book to decode a dream, or one may use a theory of dreams such as offered by Freud in order to understand a dream. The latter process is pursued in a variety of ways by a number of offshoots or modifications of the Freudian approach, and these efforts give rise to a multitude of psychoanalytic interpretations of a dream. Thus any given manifest dream may be interpreted by a particular psychoanalytic theory in order to yield a meaning, one that differs from that offered by another theory or decoding process. The facts may be the same, but the meanings may differ all the way from the particular cortical activation to an Oedipal struggle to a self-state dream. With a cautious use of the word "correct," all of these meanings are correct while quite different from one another. There is no one true meaning.

One would be not too far afield in likening the many meanings of a dream to a work such as the Bible, which likewise is considered to be a significant carrier of immutable truths that claim support from many persons while simultaneously delivering conflicting versions. Although the Bible may be thought by many to be a divine communication with but a few differences of meaning, it is also subject to a wide set of interpretations that range from that of the word of God to an interesting historical document with absolutely no religious standing. Arguments rage over each and every interpretation, and these differences change over time. So too do we see the arguments that go on between our different analytic schools with the ultimate question of whether they have anything like a "core of beliefs" or whether the differences are so extreme that they are essentially untranslatable. Inasmuch as all science is interpretable or hermeneutic, so too are any number of complex stories, musical works, political events, etc. The richness of meaning allows for the coexistence of sometimes contradictory interpretations, and it should be no surprise that any number of psychoanalytic approaches have developed over the years. This is often not only tolerated, but encouraged in the name of diversity. It should also not be surprising that diversity may lead to its own conflicts and even dangers. It is one thing to differentiate the manifest dream from the latent dream, but quite a different thing to distinguish and interpret one meaning from another.

If one serves as a reader for any one of our psychoanalytic journals, on occasion a paper is received that may present a struggle to understand, not so much due to the paper's inherent difficulty, but more so because of its unusual vocabulary or somewhat idiosyncratic theoretical terms and assumptions. One paper stands out in my mind as especially incomprehensible because of a variety of symbols and words that seem almost to resemble a secret code. A colleague explained to me the particular theory that these symbols represented, and essentially the theory made good sense once an effective translation could be accomplished. No doubt the author of this particular paper felt that he or she was writing to like-minded members of his or her group and so gave little mind to those of us who might fall outside of such membership. When Freud wrote about "the narcissism of minor differences," he indicated that groups were bound close together and so were able to keep other groups at bay by these allegiances, which were fueled by aggressive impulses (Freud, 1957, p. 199). A good deal of anger and irritation is often felt in readings of clinical cases or non-clinical papers that pursue a vocabulary of terms that are idiosyncratic and may seem meant to exclude a subset of aliens. Thus Freud's explanation seems quite correct. Yet the translation from one language to another may conceal a larger problem, since my colleague could be in error as to what the symbols and words actually meant.

It may be worthwhile to examine the origins of the many diverse groups that seem to populate psychoanalysis, as well as psychodynamic psychotherapy, along with a query as to whether this outcropping of diversity is at all beneficial or perhaps may turn out to be detrimental. Ultimately the aggression may turn to a persistent hostility between groups. Much of this is already apparent.

Most students of psychoanalysis are trained in one or another form of classical or Freudian psychoanalysis and perhaps therefore launched onto a variant of this particular classical Freudian presentation because of what are felt to be inadequacies or deficiencies in the classical approach. This initial step may be argued by some diehard classical adherents, but this is surely a common enough approach in all of science as well as psychological studies and can be evaluated only on the basis of its efficacy. This variant of classical psychoanalysis is perhaps the overwhelming explanation for much of dynamic psychotherapy, and the reasons range from pragmatism (cost, frequency, etc.) to theoretical variations (the need for a real relationship, etc.). One can argue these points as well, but one crucial component once again is that of effectiveness, not so much in the matter of symptom relief but rather of the very particular requirements of an analytic cure or effort. It would be a definitive answer to the choice of meaning if one were able to distinguish them in the very pragmatic basis of what works best. However, that seems unlikely.

The issue of effectiveness becomes murky with a change in focus to that of personalities. The important names in psychoanalysis ranging from Bion to Klein to Kohut to Lacan, with a number of perhaps less-known but equally important ones, have left a legacy of groups that are often antagonistic to one another and quite often have developed vocabularies and variations in technique that encourage a diversity that may well be more deleterious than helpful. That may well be at least partially due to the alienation that seems to be encouraged by these examples of the narcissism of minor differences. It may be the case that the initial impetus for a new "school" to be started is that of expansion of the work of psychoanalysis, but soon the personality of the founder or that of the followers becomes the overriding reason for its growth and development. Fed by popularity, a new lexicon and a need for distinction, the "school" ultimately comes to be seen as different and even as foreign by those who fall outside of its core membership. Thus a school tends to become isolated, and this isolation leads to a new set of difficulties in communication up to and including certain special words and symbols.

Distinctions

While some groups may lay claim to a particular form of effectiveness with a subset of patients, others may wish to be seen as more universally applicable and so may trumpet or merely wish to be seen as inhabiting a position that displaces and so replaces the classical Freudian approach. My own experience in a panel devoted to a case presentation discussed from varied theoretical approaches seemed to underscore the differences in the varied theories, which also allowed each to claim a certain form of self-sufficiency. Thus, from an origin aimed to be but an addition or an emendation to the core concepts, there grew a whole that was both separate and distinct. Melanie Klein's introduction of the paranoid and depressive positions was aimed at filling out some deficits in classical Freudian theory by emphasizing certain pre-genital issues that had not been fully attended to. Heinz Kohut's introduction of the selfobject concept was aimed at the delineation of certain transference configurations lacking in Freud's corpus. Yet from each of these efforts that were offered as amendments or corrections there grew groups of doctrine that became all inclusive. No doubt some members of these and other groups might feel allegiance to Freud, but it is not uncommon for members to be designated by a personality's name. If one is seen as a Kleinian or a Kohutian or a Lacanian it is assumed that one is not a member of that "other" group or of any other group. Bergmann (1993) classifies the groups as heretics, modifiers and extenders, but this classification is difficult to maintain as time sees to intensify group isolation.

If an examination is made of those distinctions that these theories seem to claim as being universal, one can construct a line of psychopathology and/or a list of diagnoses that may reveal a continuum of applicability. At one end of this line is psychodynamic psychotherapy, which can lay claim to a certain section of disorders that need not be referred to or seen by psychoanalysts and so may well be considered as not in need of psychoanalytic intervention. There is surely a good deal of room for debate about the wisdom of this segregation, but it does certainly exist. Such patients may well compose, or be included amongst or representative of, a group of patients for whom a variety of approaches either theoretical or technical can be seen as effective. Indeed, there does seem to exist a group of patients who do well with any number and kind of psychologic interaction. Thus Lacan or Kohut or Klein or Bion or Boulanger can, on occasion, guide the therapist with equally effective results. This need not mean that these schools are alike, but rather that the differences do not seem to matter.

At the other end of this imaginatively constructed spectrum are patients who may well need a very special form of analytic intervention. And here is where the arguments arise. Without dissecting the particulars of these stances, one can easily see that if a particular theoretical posture and set of assumptions can claim to be worthwhile and effective for both those cases, which may be considered the easiest as well as those seen as the most difficult, then it takes but little effort for its claim of generality and universality. Present-day psychoanalysis does not seem to be able to establish guidelines for distinguishing applicability. Although different analysts might conclude an evaluation of a patient by saying "this patient needs a Lacanian analysis" or "this patient needs a Kohutian analysis," there is surely no unanimity of decisions on such assignments. Indeed, these selections are rarely based upon a careful comparison of divergent theories but rather on what might more properly be termed prejudices. And here is where a rather glaring defect in psychoanalytic education and the entire corpus of analytic thinking is exposed.

One rather comfortable stance that exists is that of a conviction that it makes little or no difference what theory is employed inasmuch as they differ only in superficial matters. This position is defended by the claim that some patients improve no matter who is treating them, or else it is the personality of the therapist that matters without regard to his or her set of theoretical convictions. A common defense of this approach states "that it is the music rather than the words that count." Inasmuch as some psychotherapy research does support the fact that a certain percentage of patients do well regardless of treatment selection (Shedler, 2010) the same may be true of psychoanalytic endeavors.

A different form of comfort can be attributed to the conviction that there exists only one or perhaps only a few analysts who can help either a particular patient or any patient. The selection problem of deciding what works best for what patient is thereby sidestepped just as easily as the stance noted above. Both avoid the need to evaluate, compare and distinguish theories on the basis of meaningful selection. Of course this effort is easily defeated by either of the positions that insist that anything works or that hardly anything works.

Impediments

It may seem obvious that psychoanalysts should strive to practice with a maximum of knowledge in order to effectively analyze a maximum of patients. However, if a theoretical approach differs from one that a particular analyst employs, it is not surprising to see arguments arise that insist that one's own set of ideas would work just as well or better. Of course, this is a major impediment to learning anything new, and this does seem to characterize clinical presentations that are approached as deviant and different.

If one conducts a thought experiment that involves changing one's political affiliation from, say, that of a firmly held liberal Democrat to an ultra-conservative Republican, the first step involves the gathering of information about the proposed new position and the next step attempting to fit into that new and somewhat strange position. One readily available resistance to such an effort is often that of a moral one (i.e., it is wrong), and this wrong is derived from and supported by a host of moral standards that may well stand apart from the position that one is asked to consider. A similar set of propositions that seemingly stands apart from the evaluation of the proposed position is that of personal advantage; i.e., is there any personal gain to be derived from, say, becoming a Republican rather than a Democrat, and is it the morally right thing to do? Indeed, each of these questions has psychoanalytic relevance, but one often endeavors to decide upon a position as if there are independent factors operating. I vote Democratic because it is a tradition in my family or because my next-door neighbor is running for office. However, at no time can we remove these moral and personal issues, and so they direct us to more closely examine how one changes one's mind.

Differing interpretations or different meanings may at times achieve a reconciliation of sorts by an appeal to a different or perhaps higher principle, such as effectiveness or morality or breadth of explanation. The different interpretations of a Constitutional Amendment may not ever be completely resolved, but may yield to authority or moral suasion or changes in societal beliefs. However, it is unusual in psychoanalysis for a follower of one school and one personality to change to another without a period of inquiry and openness and internal struggle. More often than not such a change reflects the believer's own psychology in terms of either being an "outsider" or needing to become part of the group. And certainly if the new group seems to resonate with the person's own personal struggles, then membership is all the easier.

MacIntyre (1985) feels that any mature tradition of inquiry should be able to explain its failures (p. 398), yet a book of psychoanalytic failures that aimed to offer an explanation of failure was criticized for not presenting detailed case material to allow the readers to determine why that case had failed (Wallerstein, 2012); i.e., the intent was to place blame rather than to see how failure was inescapable within the ideas inherent in the theory. Analysts feel failed analyses are results of analytic errors or patient pathology rather than limitations of the theory.

We are faced with a choice: either each theory is immune from internal critique and thus forced to defend itself against external criticism with the inevitable result of a form of isolation, or else each theory is readily replaced by another with the result of an indifference to their differences. Philosophers have long struggled over the problem of choosing between interpretations, with conclusions ranging from pragmatism to postmodernism; i.e., pick what works best, or there just is no way of knowing or we have not yet achieved a scientific way to choose the best. Bergmann (1993) offers a test to determine what he calls "clinical usefulness," but his hope for this guiding parameter has hardly been realized. Our schools of thought have grown further apart.

Diversity and pluralism

Diversity usually carries a positive and hopeful meaning in its use. When we speak of diverse races or diverse religions we often point to the coexistence of entities that are different but whose coexistence is tolerated or encouraged. Pluralism is a bit more complex. Much like the psychoanalytic concept of over-determination, pluralism often states that a number of factors are operating to produce a particular effect. That, of course, is also different from the idea that a number of different explanations can be offered to explain a particular phenomenon. We often call that perspectivism. This raises the question of whether alternative explanations are members of the same intellectual domain. Few doubt that a particular thought or action can be seen as either a psychologic or a neurologic issue, and even within that distinction one can list both a series of causes and possible explanations. However, for the most part, psychology and neurology are separate from one another, and despite the current popularity of neuroscience they remain separate domains of inquiry.

There is a certain looseness of thinking if one collapses pluralism into diversity. Without legislating the use of language and yet remaining faithful to the Oxford dictionary, pluralism is primarily a host of causes or explanations for a given phenomenon or result, much like perspectivism, while diversity distinguishes between those phenomena or results (Webster, 1965, p. 663). The challenge for psychoanalysis is to consider whether Melanie Klein and Jacques Lacan contribute or add to a Freudian outlook or whether they are quite different ways of employing a Freudian outlook. Of course they are different, but how much does the difference matter or make a difference? The language gain here is a matter of usage, and it seems to be the case that Klein and Lacan are truly different with little hope of commonality.

Stepansky (2009) has written at length of the primarily negative effects that what he calls "theoretical pluralism" has visited upon psychoanalysis. We might term this diversity since it is really coexistence of different schools. His emphasis is one of despair: theoretical pluralism entails different theories and different treatments with no chance of achieving much more than toleration of each other in either area. Stepansky pretty much predicted the demise of psychoanalysis, and much of what he wrote seems prescient. We seem to have more and more new "schools," more and more personalities who develop new vocabularies and more and more insistence that we are all different – i.e., diverse. Stepansky scoffs at those who propose a "composite" theory of psychoanalysis with a "personal amalgamation of concepts and theories" representing a totality by noting that "one man's totality is another's abridgement" (Stepansky, 2009, p. 167).

The conclusions that confront psychoanalysis are (1) all schools (techniques and theory) do roughly the same and are equally effective, (2) different schools are truly different and some are more effective than others, (3) success in treatment depends on the person rather than the theory and the theory primarily reflects the person. This last is a variant both of (1), which diminishes the significance of the differences, and (2), which does emphasize the power of the difference, albeit primarily in its application.

The struggle over differences or diversity exists quite starkly with our psychiatric brethren who use (for say depression) one or more medications, transmagnetic stimulation, cognitive behavioral therapy, psychodynamic psychotherapy or group therapy with an equal lack of certainty as to efficacy. Sometimes it may make no difference what is chosen. Sometimes the choice of treatment is crucial. However, when one does come upon an effective treatment there may be no good reason to explain its effectiveness over others. Trial and error often rule the day.

Psychiatry is happy to admit that some medications work for some and not for others, just as any other mode may work or fail for any individual patient. Although some psychoanalysts may enjoy success with all of their patients, the majority probably have a mixed rate of success and may attribute failure to something like unanalyzability or even a particular countertransference problem. I suspect few are willing or able to take on a new technique or a new theory. Indeed, one could use any sort of example by proclaiming that Kohutians cannot conduct a Lacanian analysis or ego psychologists cannot conduct a Kohutian analysis or Lacanians cannot conduct a Kleinian analysis, etc. This "silo effect" is not peculiar to psychoanalysis or psychiatry but is an often seen product of diversity.

The danger

Some time ago a friend told me of a course offered at an analytic institute called "Deviant Schools of Psychoanalysis." The course aimed to introduce student candidates to the many non-orthodox schools that existed, but did not plan to go much beyond an introduction. Disdain and contempt are natural companions to such a course's contents, and the average student does not go much beyond the possession of a few choice code words used to describe and dismiss the deviants.

Now and then a few curious souls do pursue an investigation of a new set of beliefs such as Roy Schafer did with Melanie Klein (Schafer, 1997). In truth, one cannot easily manage a familiarity with the plethora of extant and competing schools, so the usual posture that is assumed is to be locked into the set of beliefs that metaphorically "you were born and raised with."

But consider a possibility. Perhaps the patient that seems to be at a stalemate does not require that you need supervision, but rather that a Lacanian or Kleinian approach is indicated. How would you know, inasmuch as you are locked into a position of certainty? Your dissatisfaction may be the motive, but your fundamental ignorance about other approaches is a handicap. This ignorance extends to both so-called deviant schools of psychoanalysis as well as medication, cognitive behavior therapy, etc. Indeed most analysts are not only unable to operate with a different theoretical approach but also unable to conduct a proper referral. Alas, we live in a locked-in state and talk to others and read journals with the same handicap.

Discussion

The danger in diversity derives from the borders that we construct to distinguish ourselves from like-minded groups. These borders allow us to refine our theories and techniques not only for the benefit of our patients, but also to defend intrusions from those that are stamped with the tag of alien, foreign or different. Emphasis is thereby placed on the defense. Indeed, Stepansky quotes one analyst who urges his colleagues to "be proactive in explaining and defending our view point to the public, to other minded health professionals and now, no less important, to psychoanalysts who see things differently" (Stepansky, 2009, p. 165). That time is past. We would do best to cease explaining and start listening. We suffer primarily from self-satisfaction and must work to reawaken our dissatisfaction.

At the other end of the dilemma of today is the problem of listening. Special vocabularies and closed-off convictions have led to increased isolation of our schools and have possibly driven away students. Of course, a student does not enroll at an institute that focuses on one or more "personalities" to learn about other "personalities." Some do give lip service to others by inviting representatives of these deviant groups to an occasional presentation, but these efforts are more cosmetic than substantive. In all probability this is an inherent limitation of all analytic institutes that they can only do what they do best. But perhaps one proviso or qualification should be allowed entrance, and that is one of recognizing limitations and making referrals to someone who might do a better job. Surely some Kleinian analysts should refer cases that would profit from a self-psychological approach. Surely some Kohutian analysts should recognize patients who need a Lacanian analysis. It would seem self-evident that some theories and techniques are best for some patients and not for others. Once rid of a "one size fits all" mentality we recognize both our own limitations and the advantage of other modes of practice. We thereby unlock the self-imposed state of isolation by appreciating what others have to offer rather than where others went wrong.

The process

Changing one's mind is not easily or simply explained, inasmuch as it ranges from the trivial choice on a dinner menu to a significant and meaningful choice of a lifelong relationship. However, some common features are notable, beginning with the aforementioned feeling of dissatisfaction. In a study of failed cases in psychotherapy and psychoanalysis, one striking feature that was observed early and regularly was the reluctance on the part of the therapist or analyst to admit to a failure (Goldberg, 2011a). Failure in treatment was taken to be a personal failing. In many fields, ranging from carpentry to medicine, a failure need not have such a personal connotation, and, after one approach fails, another is tried without any form of self-condemnation. Of course, if one does a bad job in a given procedure then some self-blame is warranted, but it is necessary that one differentiate the choice of procedure from its execution. However, in failed cases of analysis or psychotherapy there was usually no such distinction; i.e., the therapist failed and so needed more supervision or better training or more personal analysis. Dissatisfaction did not have far to travel until it found a home in the person of the therapist and rarely, if ever, in the practice. Thus it did not become the impetus for change as much as an indulgence in self-scrutiny. There is a somewhat automatic syllogism in the form of "if the patient does not improve then I must have done something wrong" on to "if the patient does not improve then the patient is either unanalyzable or untreatable." At times medication may be considered, but rarely does a Kohutian refer to a Lacanian. It just does not happen.

Of course the next step in a change is the move away from dissatisfaction to an alternative. Before that can occur it may be worthwhile to examine the difficulty inherent in a change that carries the burden of failure and so clearly is that of a narcissistic failure. The personalities of the various founders of these schools, with their personal vocabularies, special journals and unending efforts at proselytizing, offer a grandiosity to its practitioners which is difficult to dislodge. No one seems immune to the specialness that descends on a school that has managed to see the world in a way that is unique. It is of some interest that competing schools often jump on one or another's characteristics of this supposed uniqueness in an effort to depreciate it. Thus Lacan is ridiculed over the five-minute session while Kohut is reduced to "just be empathic." Just as the above-noted course on "deviant schools" carried depreciation in its title, so do we all attempt to raise ourselves on the backs of others. This insularity of the various schools with their needs to be different makes the next step in effecting change even more difficult.

This next step demands an accessible entrance into a difference. Unfortunately, these moves seem to involve certain tests of loyalty and allegiance, almost as if to require that one denounce one's previous club in order to join another. Once again, the crux of the problem seems to be the assumption that even though all analysis is different it is also true that all analysis is the same. We presently have no way of determining who needs a Kleinian approach versus who needs a Kohutian approach versus who needs Lacan, and on and on. It either makes no difference or it makes a world of difference. Sadly we are prisoners of our own convictions, and our convictions are prisoners of our narcissism.

The solution

A book devoted to the history of evolution describes how many of Darwin's predecessors were called infidels, and books such as *The Young Man's Guide to Infidelity* were written to warn of the traps and snares that awaited the unwary (Stott, 2012, p. xix). Something akin to this has happened in psychoanalysis with its preoccupation with deviations from the norm. No doubt we need to open our training institutes to all manner of ideas, but it seems very unlikely that a Kleinian institute will embrace a relationist approach and even more impossible to envisage that same institute to ever teach cognitive behavioral therapy. One may hope for ecumenism but doctrine is not easily transcended. Rather we are witnessing what some call "convergent evolution," which is "a story of meanderings and false starts, of outgrowths, adaptations, and atrophies, of movements backward as well as forward, of sudden jumps and accelerations and convergences" (Stott, 2012, p. 289). We are all participants in a process with a very uncertain and unknown outcome. The psychoanalysis that we all know will die and be reborn many times before a somewhat stable form of analytic therapy will emerge, one that in all likelihood will be a collection of approaches with a corresponding collection of applications.

Summary

The diverse and somewhat self-insulated schools of psychoanalysis, much like the varied modes of treatment in psychiatry, are as yet unable to determine what patients would profit most from a particular therapeutic approach. The inevitable recognition that a singular approach to all patients is unsuccessful has not led to a clear categorization of analytic limitations and capacities – i.e., when to recommend and what to recommend. Our diverse schools of psychoanalysis are more devoted to defending borders than to an open exchange of information. We need to examine the proposition that some patients profit from some techniques and not from others and so recognize what we do best. This would be a departure from the parody of boundless ability. Once we appreciate the value of other analytic efforts, we may be able to establish and utilize a common vocabulary as well as a common vision of effectiveness. That endpoint cannot be dictated but should evolve over time. The danger is a failure to recognize that we are part of an ongoing process and to assume a finality that does not exist. Perhaps if we were to recognize the danger we might accelerate a solution.

Chapter 10

Reflections on enthusiasm

In order to offer a psychoanalytic contribution to the study of the psychological state of enthusiasm, one must initially list and clarify the many uses of the term. Webster's dictionary tells us that it is a "state of impassioned emotion . . . on behalf of a cause or a subject." No matter the particulars of the feeling, enthusiasm is always for or about something or someone that is literally or figuratively outside of the person who is experiencing the emotion. We are thereby enthusiastic about an idea or an activity or even a material object. However, what may seem to be a wholesome or welcome emotion has over the years been somewhat battered in its connections with less than admirable causes, and David Hume insists that enthusiasm produces the most cruel disorders in human society in its "presumptuous boldness of character . . . especially after it rises to that height as to inspire the deluded fanatics" (Hume, 1742–1754). Thus a psychoanalytic scrutiny of the emotion should examine both its origin and expression while noting that it may have extreme manifestation along with possible breaks with reality. Not surprisingly, a careful study of the range of expressions of enthusiasm also reveals its total or near absence in some individuals who may feel themselves as mere onlookers, those who never seem to participate in the experience and excitement of this emotion – those who miss out. Thus, for some enthusiasm is a foreign emotion, for some it is ever present with no stable target and for others it entirely defines their lives and is a necessary nutrient. Whether it is normal demands some careful deliberation inasmuch as normality is itself such a problematic concept.

Development

Although babies and toddlers may show varied degrees of excitement about certain people or things, and we may well thereby attribute a complex feeling such as enthusiasm to such evidence of excitement, it is difficult for one to distinguish the two affects without bringing other issues such as admiration and value into a full appraisal of enthusiasm. Attachment studies show that insecurely attached children tend to lack self-reliance and show little enthusiasm as compared to those two-year-olds who were assessed as secure at 18 months and were enthusiastic and persistent in solving easy tasks (Karen, 1990, pp. 45). In contrast to the excitable toddler, there are also children who are listless and markedly unenthusiastic about much of the world. Therefore we may mark early development as a significant pointer to the later emergence of enthusiasm, which seems to be more clearly in evidence during latency. It is here where one can discern more obvious demonstrations of preference, of goals and values and of ideals. The formation of groups with aims as seen in clubs, the beginnings of allegiances and promotion of beliefs, along with the zeal of persuading others, often become hallmarks of early and late latency and soon become strikingly characteristic of adolescence. The fervor of team support in athletes and the devotion of belief systems in religion make adolescence the period of firming up of values, goals and ideals. And yet we cannot fail to note how certain members of this age group appear to stand outside or seem to be left out of participation in this cardinal time of "impassioned emotion."

Although it may well be true that depression is the enemy of enthusiasm, it is necessary to distinguish the listless or unenthusiastic child or adolescent from one afflicted with depression. The phenomenology of depression is complex, with some evidence of enthusiasm seeming to serve as an antidote to depression, and the absence of enthusiasm often indicating an entrance into depression. Yet some individuals who are lacking in enthusiasm need never descend into depression. It is paramount to the understanding of this psychic phenomenon that we examine the underlying construction and composition of the emotion inasmuch as surface manifestations are of little help in a comprehension of the essential meaning of enthusiasm. A series of cases will be presented, not so much for the issue of the proper treatment of this problem, but rather to present the phenomenology of enthusiasm, its probable pathologic origin and its rightful place as a distinct psychoanalytic concern as first established by Greenson (Greenson, 1962).

Clinical example: a tale of two doctors

Dr. A. came to the psychoanalyst, whom he had seen earlier in his career for help, now with what he described as a reluctance to continue his medical work. He did not so much appear to be clinically depressed as he seemed defeated and exhausted. He had first sought psychoanalysis while in medical school because of a wish to drop out of school. The analyst at that time devoted the treatment to enabling Dr. A. to successfully complete both medical training and a subsequent specialty in surgery. Dr. A. never failed to insist that he cared not a whit for his practice and felt it a continual and unrelenting burden. After that first analysis he later sought out another analyst for help with marital troubles, and it was to this second analyst that he came for his present state of an intense dislike of his work. There is no material from that first analysis available, and the ensuing clinical material will focus primarily on the phenomena of his presenting illness.

Dr. A. was a person characterized by his wife as someone who "really did not like anything." Although he accompanied her to the ballet and opera, he had no pleasure in his attendance yet did not especially dislike going. Interestingly,

Dr. A.'s approach to most activities did not emphasize either strong positive or strong negative feelings. Life was endured with a minimum of caring. The present illness was said to be depression, according to Dr. A., who said that he was depressed and felt separate and distant from most of the issues of his life. He was not sad, not suicidal, not obsessed by dark or dismal thoughts. Rather, he felt disconnected from his work and from the people in his life. Work was not only burdensome but also caused him a good deal of anxiety, because he feared making mistakes and so being exposed as incompetent and ill-prepared. He certainly did not wish to become more involved or interested in anyone or anything. If pressed he would admit only to a wish to return to the state of mind in which things did not matter much at all. He recognized that something was missing in his life, and a regular involvement in gambling had somewhat relieved this supposed deficit. However, even his gambling was experienced from afar with no eagerness or excitement, as he might follow an athletic event on which he had placed a bet. In this respect he was much like his father who, although a cardiac invalid, played cards and was ever credited with his "poker face." This identification with his father as an occasional gambler was kept hidden from his family and friends and appeared to have no central role in his life. In order to better elucidate the underlying structural deficits that are so characteristic of a failure of enthusiasm I shall turn to a detailed report of a previously described case (Goldberg, 1975).

This patient, whom I shall call Dr. B., was a divorced, 40-year-old physician, the father of three. He had likewise had a previous analysis. The present and previous treatments were initiated by his being discovered in his perverse behavior. This took the form of fellatio performed on the patient by female patients whom he stimulated during routine physician examinations. The number and variety of such responses was unusually high, and the patient rarely repeated such behavior with any single patient. I think this does classify as a perversion since this was an episodic activity associated with anxiety and followed by what he described as guilt but which often seemed clearly to be shame. The willing participants were unnamed and unknown and hardly more than vehicles for the act.

The patient's past history was that of a poor boy from a large family of six children raised in a small town in Canada by an extremely cold and religious mother and an alcoholic father. He recalls his mother as being a demanding but unresponsive woman for whom expression of emotion was a sign of weakness. Father became ill when the patient was a teenager and stopped working until he died several years before the patient entered analysis. The patient was unusually close to a brother who died in an accident when a young man. The patient was markedly moved when discussing this. An outstanding feature of this patient's childhood was an undiagnosed illness which began when he was eight and continued until he was 16. He had repeated intense pain in his right femur which occurred for several days and nights and then would subside for an undetermined time. The doctor could find nothing wrong with him and finally accused him of just not wanting to go to school. Thus the patient would endure this pain silently by himself night after night. He would often awake and sit in his mother's chair and listen to the radio and/or masturbate. Interestingly, no other member of the household was disturbed during the night, and the next morning the patient felt unable to tell his parents what he had endured the previous night. Finally, at age 16, his older brother took him to an orthopedic surgeon who diagnosed a chronic infection of the bone and subsequently operated with complete success.

The patient had begun his first analysis shortly after his perverse behavior was first discovered. He reported that his deviant activities began after the birth of his first child, a boy. That analysis had concentrated on the perversion as a manifestation of a superego deficit, and after some time the activity stopped and analysis terminated. The patient felt much better during this analysis and felt sorry that his symptom returned rather soon after termination. He was hesitant to disappoint his analyst by reporting such an exacerbation.

The present analysis began after his perverse behavior had been going on for several years. The patient reported it with extreme tension and agitation, and at times he was inaudible. Again, the analytic effort will be condensed for purposes of illustration. The patient soon settled into a stable transference wherein he longed to be a part of a strong, secure imago. The fellatio behavior seemed to illustrate the connection of a weak person to a powerful one. The woman of the behavior was represented in dreams as an old woman and seemed to be the weak, dependent image of the patient. He struggled to be the strong man with the big penis as well. The perversion diminished to a point of an occasional recurrence, which was seen as representing disruptions in the analytic situation. On one occasion the patient's son became ill and needed surgery, and the patient cancelled his hour in anticipation of the procedure. An outbreak of his perverse acting out resulted, and the dreams and associations revealed his struggle over his wish to ask the analyst to be available in case he was needed. The patient had a tremendous conflict over wanting to call the analyst just to let him know what the surgery revealed. He then recalled his overwhelming longing to tell his parents at the breakfast table that he had been in terrible pain at night but clearly seeing in his mother's face that he had best not complain. He next remembered a tonsillectomy when he was four years old. He was ready to leave the hospital, and his father picked him up and carried him out. He had remembered this previously with a feeling of outrage at being treated like a baby when he was old enough to walk out like a big boy. But now, for the first time, he recalled how marvelous he had felt to be held securely in his father's arms and how he had longed for such a union with a powerful person. The perversion was now to be understood as mobilizing this very feeling of longing for union with an idealized and omnipotent figure and as becoming rapidly and effectively sexualized. The longing and the structural need became sexualized and in the behavior one sees not only the sexualization of the idealized parental imago and of the grandiose self, but the feeling state as well as experienced sexually. Once the fragile self had become more structurally sound, Dr. B. began to pursue the medical specialty that he had long admired from afar. He experienced an enthusiastic participation in work that he had previously felt was not available to him, and he thus demonstrated the missing component of enthusiasm. The formation of a well-structured self is posited on the development of an ideal self derived from the internalization of a lasting experience with an idealized other.

A tale of two artists

Ms. A. was an established artist who had entered psychoanalysis for severe hypochondriacal preoccupation. She was constantly in search of reassurance about a wide variety of ailments and was able to pursue her artistic work only when free of worry about her health. She continued psychotherapy after her analysis and seemed unable to participate in termination, despite the radical diminution of her physical concerns. Her past history was significant in that her mother was chronically ill and was admitted to a nursing home during Ms. A.'s teens, an event that precipitated a withdrawal from school and a lifelong personal regret at not graduating from college. In the psychotherapy that followed the analysis, it soon became clear that her efforts in pursuing her art essentially alternated with her hypochondriasis and thereby evidenced an enthusiasm that also came and went with her work. She could be amazingly excited about a particular work of art, which she shared in detail with her therapist, but would inexplicably become remote and distant when she stopped her work. The initiation into periods of excitement followed by fallow periods of lack of interest and physician concern soon became clear as a product of disruptions in the transference. The therapist soon became involved in an enthusiasm over the patient's works of art, and this shared enthusiasm could be interpreted in terms of the patient's mother's illness. The hypochondria over time became an object of ridicule by the patient.

In contrast to Ms. A., Mr. B. never lost his enthusiasm for his artistic work. He also rarely lacked for inspiration and needed no impetus to begin work on a new project. He had not begun his work as a sculptor until late in life and wondered if this late start was because his mother was such a talented and prolific artist who often said to her son: "I am not only your mother, I am an artist." It was not the case that his mother in any way discouraged his artistic efforts, but more or less it seemed that art was her domain. When he did begin to devote himself to art, he did so with undiminished enthusiasm, save for a never fully satisfied wish for others to recognize and acknowledge its value. He especially wanted his therapist to view and appreciate what he had done. An interesting countertransference reaction to the patient was a wish to buy, own and enjoy a work of the patient. Now it goes without saying that most, if not all, artists, save perhaps for Van Gogh, want an acknowledgment and appreciation of their efforts, but Mr. B. seemed to dwell on a failure of proper and sufficient recognition. However, this lack of public response never seemed to diminish the joy in its execution. Whenever Mr. B. entered his studio, he was caught up in an eagerness for work. His enthusiasm was undiminished. In that area of his self he had succeeded in uniting with his idealization of his mother and her devotion to art, but, perhaps unfortunately, he remained longing for a proper mirroring of his own efforts as was so evident in his therapist's struggle to fulfill that yearning.

Enthusiasm and meaning

There is probably no brain localization for enthusiasm, since it is likely composed of a network of feeling that gives meaning to much of our lives. It can range from a quiet feeling of satisfaction over a moment of exchange with another person to a loud and vociferous exclamation as a part of a group. Either way it registers our union with something or someone that means something to us that we admire and value. Some are satisfied and enthusiastic over a solitary pastime such as stamp collecting, some are satisfied and enthusiastic over the achievements of an athletic team. They mean different things but they essentially mean the same. Enthusiasm gives purpose to our lives and it is the merger with persons and ideas that carries us along and that allows that vital union. One collects stamps as a solitary activity, and it may have a detached, almost robotic, feeling, or it may allow one to become a part of a larger community of like-minded collectors. One can watch a football game with no concern as to winners or losers or as a greater group of fans for whom the game has great meaning. The meaning conveyed on the action comes from the union with a feeling of purpose that is the primary attribution of enthusiasm. We thus enter the arena of union of cause and motivation and leave the concern with facts and evidence. People and ideas mean something to us.

Enthusiasm and excitement

If one pursues the lack of enthusiasm as being essentially due to a structural deficit in the pole or area of idealization, then it may well follow that another form of fragility can be seen as responsible for a failure in allowing an associated experience of excitation along with the urge to express and promulgate one's personal values and goals. One may well be in possession of such values and goals which serve to guide a course of life and allow for a sense of certainty and solidity yet which are unable to be expressed or extended to others. As with Dr. B. the excitement itself requires structure. Having a belief system and being openly proud of one's belief system are not necessarily coupled phenomena. We often see analytic candidates who declare a fervent belief in psychoanalysis but are ill equipped to exactly say what they believe. These students may gradually learn and incorporate the ideas of analysis and become excited by these ideas. The task that then presents itself is the marriage of what they know to a more genuine set of goals and values which can lend direction to their lives. One must live what one believes in order to be enthusiastic about it, and the failure to do so is readily seen in the cults of personality.

Enthusiasm and the person

We are all, sadly, too familiar with a set of goals and values tied to a person, along with the ensuing glorification of that person. From Martin Luther to Melanie Klein, we are all aware of the introduction of new ideas which are often at odds with an existing set of beliefs. Usually these breaks with the extant system are a beginning addition and/or a modification of what is the reigning system and are necessary, albeit difficult, to digest. All too often the changes become linked to the person or persons who are champions of this new vision, and the newer set of ideas may develop into a cult. We see this in certain religions or in any set of ideas involving goals and values which call for admiration as well as proselytizing. The difficult part of what should ultimately be a gradual process of modification demands our going beyond the person. This problem is quite representative of adolescence, a time of strong emotional ties to ideals who are often athletes, sometimes teachers and rarely entertainers. It is the movement of concepts that go to a level of thinking, which we might call depersonalization, that is needed. A good example of this can be seen in mathematics where Euclidean geometry became just geometry and Newtonian calculus became merely calculus. Growth requires relinquishing the personalities. It need not insist on the name itself being dismissed, but the name cannot serve as an inhibition to change. Idealization of the personality can inhibit an enthusiasm about ideas or beliefs.

Enthusiasm and morality

Most systems of beliefs, as well as collections of values and goals, serve as limiting or regulating factors in behavior. Issues of right and wrong, good or bad, allowable or forbidden become carriers of and reflections of the value system. In psychoanalysis the boundaries of behavior regularly reflect the dominant theory. If insight is a goal and interpretation is the method, then it follows that suggestion is seen as wrong. We may admire and foster restraint while we abhor any action on the part of the analyst as being nonanalytic and so non-therapeutic. We struggle over deciding whether something is inherently wrong or is wrong only because it falls outside the theory being employed. Thus we do not hug a patient or even shake hands or offer comfort because it may be seen as behavior that is theoretically unwise. So, too, do we see some behavior as wrong in and of itself and proceed to weave it into our theoretical stance. For instance, one may feel it morally correct to be honest, yet telling a lie to a patient may not, in some instances, make much of a difference (Goldberg, 2007). It is good to be kind and thoughtful but that is best seen as a universal virtue that need have no special relevance to an analytic practice; i.e., many moral and ethical standards fall outside of psychoanalytic ones, and many psychoanalytic standards may fall outside of what are considered normal interactions of people. Devotion and admiration to a belief system easily slips into rules of behavior that may or may not be relevant. Our enthusiasm can sometimes blind us to questioning our beliefs and values.

Enthusiasm and change

Many people define themselves by one or more belief systems, as in the description of being a "Christian Democrat pianist specializing in the music of Bach." Each of these elements of description reflects a modicum of pride in what one is and an idealization of what one values. In order to substitute one ideal for another, say deciding to vote Republican, what is required may call for a reorganization of the self. Rebuilding is a wrenching experience that may well reflect the degree to which your belief system literally holds you together. It is a valuable exercise in the consideration of psychopathology to distinguish between the capacity to be enthusiastic from that of the non-flexible need of a particular form of external ideals. Of course, the capacity for enthusiasm per se has its own particular set of ideals and must join with an acceptable set of goals and values. One may become a Republican because one values the substance of that party over that of the Democrats. The attempted change from one value system to another may occasionally result in feelings of emptiness and alienation, and for some it is an impossible movement. In psychoanalysis we cannot fail to note how certain theoretical systems allow us to feel connected to others, as well as integrated. If, however, one moves to another system of theory and practice, a wholly different and more dominant set of ideals, such as being more therapeutically effective, must be introduced to facilitate such a change.

We need to recognize a hierarchy of ideals with some that are readily interchangeable and some that seem to be more bedrock. It is a rare individual who can examine his or her beliefs, unless that very act of self-scrutiny occupies a higher status in one's personal hierarchy.

Although it may seem easy to change an allegiance from one baseball team to another, no such ease of transition could apply to religions or even psychoanalytic theories. We are regularly blind to our beliefs in terms of justification unless and until we are called upon to defend them. Such a defense demands a higher set of beliefs, as when one claims that Lacan helps more patients or that Melanie Klein is more honest about unconscious fantasy or that Kohut is more sensitive to patients' needs. While we may consider such differences between our theories, we are rarely open to careful comparisons between sets of ideas because we so need to believe what we believe in.

Discussion

When a baseball team wins a World Series the fans may become so excited that they break windows, overturn cars, start fires and in general wreak havoc. We see this form of fervor in wars, in elections, in religions, and we puzzle over its intensity until we recognize the unconscious wish for union with an all-powerful and all-knowing ideal, something our parents were and needed to be for us. The manner in which these early ideals were internalized and organized into the structures of our selves determines our future capacity to be enthusiastic about all manner of things, from baseball teams to automobiles to Nobel Prize winners. Enthusiasm is a normal and necessary emotion, and the extremes of its realization range from near total absence to uncontrollable excitement. And so it deserves a place in any history of any patient. One must determine both its absence or presence along with its frequency. Some patients use it to ward off depression, and it often takes the form of a zealous participation in causes. Patients who are never without a cause, who move from the pursuit of one "cause" to another, find their lives empty without a cause and become enlivened only as part of a greater cause or purpose. The nobility of their participation in a cause may serve to disguise a hidden depression and all too often remains unquestioned. An investigation of enthusiasm should not be considered an exercise in armchair psychoanalysis, but rather as a necessary part of all analytic inquiries.

The psychopathology of enthusiasm

The study of enthusiasm allows one to construct a continuum from "over enthusiastic" to "lack of or absence of enthusiasm." There are those who seem always to seek projects or purposes to join, as if they are in constant need of stimulation. That is one end of the continuum. There are also those who seem quite unable to become involved in any form of project or cause. They stand outside of any situation that may require a modicum of passion. That is the other end of the continuum. Essentially it seems fair to say that a state of impassioned emotion is equivalent to feeling alive while its absence reflects something akin to deadness.

The good fortune of enthusiastic individuals who are never at a loss for projects or causes of interest is not that of those who literally depend on something or someone else to enliven them. They succeed in their quest at times with a minimum of regard for the particular project, inasmuch as the need for joining a cause may outweigh the essential value of the cause. Indeed, a careful evaluation of those who "go whole hog" for anything at all reveals how vital the pursuit of purpose serves to keep at bay the very feeling of deadness that characterizes the supposed opposite end of the continuum. Perhaps one can safely presume that the problem common to both extremes of enthusiasm is that of the very feeling of "deadness."

Too often the search for "feeling alive," and so combating the feeling of deadness, is categorized as a form of depression. However, a careful diagnostic evaluation of such patients need not reveal any of the usual signs of depression, such as problems in sleeping or eating, thoughts of suicide, feelings of sadness or preoccupation with morbid or dark thoughts. Lack of interest is simply not the same as sadness, and our psychiatric brethren do not seem presently capable of such a careful delineation.[1] Often stimulants are added to the medicinal treatment of depression, but are not well handled by anti-depressive medication. These stimulants allow for a continual combating of the dread of deadness by producing states of excitement.

The psychopathology of enthusiasm, which reflects a lack of purpose, may be a lifelong problem in patients whose self-structure does not enable them to have properly internalized a strong set of ideals, as well as those whose lives are spent in a continual pursuit of external ideals. Neither of these conditions are reflective of the usual dynamics attributed to depression, although their superficial appearance may suggest depression. One can readily see the distinction in those individuals who, after the successful completion of a major project or after retirement from a productive life, complain of feeling lost or empty and who subsequently are quickly categorized as depressed. Of course they may well be phenomenologically depressed, but a new project is readily recognized as the best of treatments. Those individuals who are able to remain enthusiastic and move with care to new projects reflect a well-structured self-organization with strong goals and values. Those individuals who remain primarily filled with regrets over what they lost or finished ever remain dependent on external factors to properly enliven them.

A psychoanalytic contribution to the study of enthusiasm might serve to establish a diagnostic category based less on overt phenomena and more on the development of lasting ideals reflected in enduring goals and values. The ease by which a feeling of lifelessness is read as equivalent to depression may do a disservice to patients who may profit more from psychologic than pharmacologic treatment. Many depressed patients do not have such problems as reflected in those with lack of enthusiasm and so demonstrate a clear difference between the two types of affliction.

Clinical example

Mr. M. was a 65-year-old, successful business executive, the retired head of a manufacturing company, and lost his wife of 40 years and found his life empty and meaningless. He went to a psychiatrist who placed him on anti-depressive medication to no avail, despite a series of dosage and medication changes. He was given a series of transcranial magnetic stimulations, again with no success, as well as a course of CBT therapy. Finally, as an act of last resort, he was sent for analytically oriented psychotherapy.

Mr. M. was an only child who had a doting and overprotective mother and an extremely distant father who was said to be a very important person in the community. Mr. M. could recall the few moments of care and concern by his father, who seemed to be a very busy person with little time for his wife or son. Mr. M. had few interests or ambitions for himself until his world suddenly changed with the entrance of his future wife into his life. They met in school, and the marriage that soon followed was reported as an unadulterated good fortune. His life became filled with purpose and meaning, with children and grandchildren as continuing evidence of his satisfaction and lasting adoration of his wife. Her death brought little alteration in his life, save for the emptiness and lack of purpose that he complained of to his varied therapists. Interestingly enough, but not at all surprising, Mr. M. was seen as a case of unsuccessful mourning, although he did not have any overt signs of sadness, tearfulness, suicidal ideation or vegetative symptoms. He hated to be alone and was rather happy when he had plans to do things with others. There was no evidence of ambivalent feelings toward his deceased wife, no material relating to anger, not so much a feeling of loss a that of a loss of purpose. Shortly after starting the psychotherapy he decided to quit because it was no better than medication. The therapist could only feel that he had missed something, and only much later did he wonder if this was a misdiagnosed case of a lack of enthusiasm which had been offered by the wife when she was alive. It dated back to the search for ideals which had been a result of the failure of the union with his father, and which had been solved by the connection with his wife who offered him purpose. He now needed not to grieve for her loss but rather to regain some meaning to his life.

A variant of enthusiasm

Dr. K. came for analysis because of a variety of problems involving work, finding someone to marry and a persistent longing for a relationship with an older man. He described his childhood in terms of his being the favorite of two boys but yet of frustration at being kept away from his father by his mother. Father was ever busy writing textbooks for his classes, and it was clear to everyone in the family that there was a need for the extra income derived from these books. Father was not to be disturbed while he was shut off in his study, while the mother, who was herself an anxious woman, was surely no substitute companion. Dr. K. became closer to his now available father in his later years, but the father's abrupt death left Dr. K. feeling unfulfilled.

At the termination of his analysis Dr. K., now a married father of two, did have what one might consider a symptom involving enthusiasm, while others might consider it a normal variant of enthusiasm. Without going into the details of the analysis, which involved the development and working through of an idealizing transference (Kohut, 1984), it may be profitable to evaluate this persistent symptom against the background of the pathology of enthusiasm. This supposed remaining problem consisted of a periodic enthusiasm and excitement about new ideas or new inventions or new investments. Dr. K. and his wife stumbled upon a newly opened restaurant that specialized in certain original, ethnic dishes, and Dr. K. spoke to the owner with a fervent wish to become a part owner and investor, because he was so excited about the food. These episodic and periodic bursts of enthusiasm were never unrealistic and often were quite productive, but in some manner they appeared to be quite necessary. Dr. K. terminated his treatment with their persistence. It would appear to be the case that periodic moments of enthusiasm, be they about new candidates for elected office or new investments for financial gain or even new tastes in food, may all fall outside of the pathology of enthusiasm.

Summary

The study of enthusiasm is an overlooked step in some psychoanalytic evaluations inasmuch as the positive aspects of enthusiasm, both in its role in worthwhile causes as well as its welcome by most people, are felt to be more a sign of health than illness. There may be a natural resistance to subject it to psychoanalytic scrutiny. If we see it as an indicator of the development and ultimate integration of psychological ideals, it can then be recognized as extending along a developmental line, beginning with an attachment to powerful figures as external representatives of idealization and proceeding to a mature, internalized set of goals and values. Along this developmental line can be seen evidence of perverse activity as a sexualization of the defective structure, as well as a lifelong absence of enthusiasm. Perhaps no history is complete without a proper study of this common "state of impassioned emotion." It is a facet of human personality that must not be overlooked in evaluating psychological health and illness.

Note

1 The diagnostic manual (DSM-IV-TR) lists the symptoms for depression and does include diminished interest or pleasure but primarily concerns itself with issues such as weight loss, insomnia, fatigue and suicidal ideation. The feeling of deadness might fall under the symptom list of feeling worthless or diminished ability to concentrate (American Psychiatric Association, 2000, pp. 168–169).

Chapter 11

On outrage and the need to be mad

Introduction

Although further studies in neuroscience may well succeed in determining, with a good deal of exactness, where activity in certain areas of the brain reflect expressions of anger and rage, the particular meanings of the rage and anger may not be so revealed. That is to say that the distinctions between the mind and the brain may remain, and this may well be true of the unconscious psychological contributions to the emotion that is being experienced. One may be angry at the car that cut you off at the highway and so be a victim of or a participant in "road rage," but the connection to your being tripped up in a soccer game when you were about to score a goal at age seven and/or the further connection to your baby brother usurping your "only child" status when you were age two may not be apparent or even readily accessible. Of course, one may feel that there is no need to pursue rage much beyond its appearance and its neurological correlation, but the choice beyond eradication versus complete causal understanding – i.e., between physiology and meaning – is a stark one and one that may not be based on ease of application.

One immediate need for the differentiation between the brain site or sites and the psychological derivatives is provided by the fact that a similar expression of rage may reflect quite different psychological meanings, as well as the possibility that different psychological meanings may exhibit different expressions. If there exists a one-to-one correlation, there is little or no problem. But diagnosis may be remarkably different if these correlations do not hold. What at first glance appears to be just "rage" might possibly be subject to further, more careful categorization. We have learned from "depression" that it is not merely a manifestation of a certain set of phenomena, since antidepressants work on some depressions and not others. Perhaps this collection of diagnoses is better understood as not so much how it looks but rather what it means. Therefore it may well make a difference if the rage that is expressed is not merely that of one brain area being activated but rather is best understood by investigating what it means, with the clear recognition that it may mean different things to different people, as well as different things to the same person at different times.

The psychology of rage

It, at first, may seem both challenging as well as discomforting to think of rage as having both a positive as well as negative side. We often, almost automatically, consider rage to be an exemplar of a loss of control which may inevitably lead to guilt, regret and remorse. We are sorry that we got so angry, perhaps more so that we expressed it. Yet there seem to be those for whom rage is a constant companion with but rare and infrequent respites. There is a wonderful cartoon by William Steig of a little man inside of a box with the caption: "I do not forget to be angry" that so exemplifies the fact that for some people it is very necessary, even vital, to remain angry and not to let it lapse.

The psychological explanatory antecedents to this need for rage range from it being a reaction to an equally large spectrum of misdeeds. Often it is difficult to tease apart what is considered to be normal aggression from justifiable anger from blind rage. For some there may be no meaningful difference since they constantly are plagued with rage no matter what the context. For others, they may feel almost incapable of getting very angry about almost anything. It is clear that an explanation of rage, its causes and its place both in our psyches as well as in our society is in order. Heinz Kohut differentiated aggression directed at objects from those involving narcissistic rage on the basis of their subsiding when one reaches one's goal (Kohut, 1984, p. 138). However, there may also be a difference in the very experience of the aggression, and for this distinction we must examine the phenomenology.

Phenomenology

The conscious, more overt manifestations of rage must begin with the recognition that it is experienced as pleasurable by some and often has such an erotic component that it is sought after in a variety of ways. Although it may be relegated to being but an expression of a drive, the erotic aspect that may accompany it can often be described as enlivening and invigorating. Indeed, the pleasure that accompanies expressions of rage may lead to it having an addictive quality and may seem for some that they need to be regularly angry in order to feel real.

Case example

Dr. F. came to see a psychiatrist upon the insistence of his wife, with whom he fought almost constantly. As the therapist listened to the present-day report of the marriage, it was indeed punctuated by regular and repeated squabbles over what seemed to be minor, or at times manufactured, issues. Dr. F. could hardly be blamed for instigating these battles (for they surely were rightly called battles) since both he and his wife came across as full and eager participants. The regularity and the almost trivial initiators of these marital arguments led to the therapist's asking if they were almost welcomed by the participants who seemed to both expect and even enjoy them. Dr. F. seemed unable to deny this possibility, but initially could not admit it.

When Dr. F. would reflect on his personal appraisal of his own and his wife's mutual rage, he could admit to a certain readiness to get angry, along with a relief at its expression. He could reluctantly recognize the lift that rage offered him as well as a hidden anticipation of their next fight. After a good deal of therapy, Dr. F. could also experience and talk about his missing the fighting and feeling a certain sadness at its unavailability and possible permanent loss.

It does seem fair to consider rage and rage attacks (i.e., the chronic and the acute) as having both erotic and addictive components. Inasmuch as these are conscious factors, albeit often only reluctantly acknowledged, it is necessary to investigate the unconscious factors that operate in expressions of rage. A psychoanalytic study of rage is not usually available in the extreme forms of this emotion, especially those which lead to physical abuse and/or bodily harm, and such cases are routinely considered primarily as exhibiting loss of control and/or primitive defenses and immature ego development. However, many individuals suffer a good deal from their struggle with rage, and although they can be successful in its suppression many are unable to find relief from its ever-present pressure to be realized. They cannot forget to be angry. Dr. F. may allow us an entry into an explanation of his rage by our utilizing Kohut's ideas about a good marriage (Kohut, 1984, p. 220 [note 11]), which allows one partner to provide a necessary selfobject function for the other temporarily impaired self.

A psychoanalytic theory of rage

Although the basic orientation and vocabulary used in this explanation are those of self psychology, no doubt another lexicon and theoretical frame might easily be substituted. The theory is best approached by using some of the salient points of Dr. F.'s treatment, which was replicated in several other analytic experiences with angry patients.

Dr. F. had had a number of unsuccessful treatments with other psychiatrists as well as with a variety of medications, but he soon settled in to a somewhat comfortable therapeutic relationship characterized by recurrent wishes to argue with his therapist. Soon, however, he looked forward to treatment along with a wish to increase the frequency of his sessions, a wish which he seemed unable to quite make happen. As his fighting with his wife began to diminish (much to the unhappiness of his wife, who complained of his wasting his money on treatment) and as he often recognized an inner feeling of emptiness, he became periodically enraged and suicidal. In fact, he said that he was upset enough to kill himself, but he did not appear depressed or develop any plans to do away with himself. He was more aptly "mad enough to kill himself." This madness may be seen as a beginning self-fragmentation which could best be relieved by any manner of anger which seemed to serve to solidify himself. However, Dr. F. showed no signs of fragmenting. Thus he appeared to demonstrate the fact that he had a very fragile connection to those selfobjects who served to offer him support and integration. As he lost these fragile connections he resorted to anger to both bolster his self-structure as well as to connect with the selfobjects which had but a tenuous relationship with him. In the transference he struggled to fight with his analyst or to free himself from his analyst, neither of which positions offered him the stability of a firm self-structure that he longed for. He needed others but could connect only with anger, and harbored a wish not to need them.

Dr. F.'s anger was a double-edged sword, since it was deployed in order to make him feel alive and integrated, but its appearance and overt manifestation may have been a forerunner of self-fragmentation and often increased rage. Over time, as he began a relationship with a new woman along with a divorce from his wife (who continued to attempt to engage him in a variety of endless arguments), he began to form a more stable and integrated self-structure. Dr. F.'s life and early development had ever been one of unfulfillment, with almost every aspect of his life somehow falling short of what might have been. His major identifications were with his chronically discontented mother, who could not or would not ever be satisfied with her husband, her children and her life. Nothing made her happy, and Dr. F. never gave up trying inasmuch as he needed her to better form a self of stability. She was unreliable but needed.

Some patients who live with rage may restrict it to circumscribed areas such as Dr. F. did at home with his family, while some extend it to their entire existence. The degree and the area involved can reveal a portrait of the self, its overall structural strength and the places of weakness and possible fragmentation.

Case example

Mr. K. was a lawyer of some status who has recently been asked to step down as the managing partner of his firm because of a diagnosis of presenile dementia. He was referred to a psychiatrist by his neurologist because of a change in his personality from a soft-spoken, competent lawyer to a very angry man given to rageful outbursts. The interview with the psychiatrist seemed to show that Mr. K. was very reasonable about limiting his practice because of his newly diagnosed impairment and had agreed to relinquish his position as managing partner. However, he became incensed at the suggestion that he give up his practice entirely, a new decision which had been an idea of some of his colleagues. He felt entirely capable of continuing to service his clients despite some minor memory lapses and an occasional struggle to concentrate. In fact, he felt that he was as efficient and able as ever in spite of what he felt were rumors circulating around that he was "losing it." Nothing upset him and enraged him as much as this conviction of the lack of trust of his fellow lawyers. He wondered if they were planning to spirit his clients away as well. He confessed to an almost daily struggle to feel comfortable in a practice that had always been a haven of comfort for him.

K. was feeling a slow disintegration of his self and a gradual withdrawal of those selfobjects which had heretofore sustained and maintained him. His rage represented both an effort to re-establish his self as well as to forestall a fragmentation of the self. It could also be seen as an expectable reaction to what he viewed as mistreatment, but one wonders if that explanation is a construction designed to make sense of irrational behavior. K. also illustrated the patchiness of a self-disorder inasmuch as the rage seemed confined to the failure of mirroring and validating selfobjects; i.e., "you are as good as ever; you are your old self" as opposed to "you are not what you used to be or what you should be." The latter message led to self-disintegration and the ensuing anger. K. was able to admit some of his cognitive failure but not able to do the same with his emotional incapacity. Denial of his cognitive incapacity served to keep a certain level of self-esteem intact inasmuch as confronting it might well have devastated him.

Selfobject failures are ubiquitous, universal and ever ongoing. Most of us are able to readily exchange selfobjects and so maintain our self-esteem and structural stability. If our feelings are hurt by a form of narcissistic injury we may feel a temporary depression or we may react with anger or some other emotion, but over time we all have devised mechanisms of recovery. Something quite different occurs when the core of the self begins to fragment. One might contrast the decline of K. to that of Miss M., who was diagnosed with early onset Alzheimer's disease when she retired form her job as a social worker. She accepted the diagnosis, made plans for future caretaking and slowly became more and more unable to manage her life. Although some may have thought her to be depressed, she seemed primarily resigned to her impending fate and never showed any sign of anger or resentment. She remained herself to the very end. Her self-structure had an inherent stability that K. did not possess.

Discussion

The challenge that presents itself is that of forming a clear differentiation between patients who have an unsatisfactory nuclear self which is ever on the verge of fragmentation – i.e., those who become enraged at the failure of their selfobjects to support, sustain and aid in self-integration – from those who require their selfobjects to maintain their self-esteem either in one or another aspects of the self – i.e., those who are hurt by a selfobject loss or any other form of narcissistic injury but are still able to move to others for sustenance and integration.

First we must decide if the distinction between patients who struggle with fragmentation is valid. Dr. F., at least, showed no evidence of what might be considered a borderline personality disorder nor did he ever have any hint of psychosis. He did, however, seem to restrict his rage to a particular form of narcissistic injury having to do with his presence and his efforts being unrecognized and dismissed. Although a large sample is needed to examine and validate the hypothesis, it initially seems true that rage is not readily correlated with the severity of pathology. All manner of people become angry, and it is only in the expression of anger that we seem able to differentiate degrees of mental illness.

It does, however, seem possible to isolate a particular dynamic that ushers in rage and allows it to flourish. With full recognition that only particular areas of the brain are activated and noting the endorphin-dopamine interplay, we turn to a psychological narrative for a dynamic explanation. Umberto Eco, in his fascinating essay "Inventing the Enemy" (Eco, 2013), tells us that making enemies is a continual and relentless occupation (p. xi). He goes on to warn us that efforts to understand those whom we hate by invoking a sense of morality renders us impotent and helpless. Eco quotes George Orwell's *1984* rendition of the induction and perpetuation of rage in the group involved in what was termed the "Two Minute Hate," a mechanism to rouse people to a frenzy. He states that "a hideous ecstasy of fear and vindictiveness, a desire to kill, to torture, to smash faces in with a sledgehammer, seemed to flow through the whole group of people like an electric current, turning one even against one's will into a grimacing, screaming lunatic" was what everyone joined in and became (Orwell, 1993).

Perhaps we can employ this necessary construction of an enemy or of someone different from ourself by joining Eco in his insistence that having an enemy helps us to define ourselves by providing an obstacle against which to measure our own worth. Although Eco supports his thesis by recounting the history of enemies and hate from Saint Augustine to Hitler, we can readily translate this to our own comprehension of how anger and rage allow for a feeling of self-definition and self-stability and so also how understanding and empathy can quickly remove that. In *1984*, the Two Minute Hate was a necessary and repetitive activity planned for group cohesion. If we transpose this to our daily life we can distinguish individuals who become enraged only when their self-delineation is threatened by some sort of narcissistic injury from those who need a regular source of hate to remain well defined, from those for whom rage may be destructive rather than enhancing a feeling of self-definition. Like an addicting drug, some can never stay a way from it, while some must avoid it at all costs. And like an addicting drug it can hijack the neural system and become needed for survival.

If we revisit "road rage," for some it may be explained as the disruption of a self-stability achieved by cruising along and feeling in control as one drives on the highway. This stable and integrated feeling is disorganized when your car is cut off, and the need to re-establish the self is accomplished by and with an angry outburst at the anonymous, offending vehicle. One, of course, may enlarge on that explanation with the inclusion of specific childhood antecedents, all the while keeping in mind that not everyone becomes enraged by being cut off, and therefore the very idea of raising it to a diagnostic category seems to be a very superficial exercise.

Empathy as the antidote to rage

If one hates Adolf Hitler and all of the awful things he stood for and the atrocities he committed, then there is surely no good reason to put yourself in his shoes and so attempt to understand him. We have the "other," and the difference sustains and maintains the hate. Once we step across the divide to become empathic with the "other," the difference becomes obliterated and the hate may be dissipated. As his therapist listened to Dr. F. describe the latest fight with his wife, much to his dismay the therapist might attempt to explain the wife's behavior and so allow him to "understand" her. Dr. F. was interested only in "using" the explanation as ammunition to fuel the fighting rather than as a way to defuse the situation. So, too, can we "diagnose" Hitler in order to better explain what a monster he was, but never to go so far as to forgive him. It is an interesting exercise in the application and extension of empathy to note how it erodes one's antipathy toward what is initially seen as foreign and bad. One can employ empathy in order to make use of it, as seen in certain salespeople and perhaps in psychopathic personalities. In this manner, one can preserve one's disdain and contempt for another, but if a moral dimension is added to the effort then hate and anger may be lost. To the degree that rage allows for strengthening and stabilizing the self, it is not easily surrendered. Some people need to be angry, and that need must be recognized and appreciated before undertaking its removal.

Summary

Both love and hate are powerful emotions that bind people together, and Freud's pointer to "the narcissism of minor differences" (Freud, 1930) alerts us to the crucial role that differences play in instigating aggression and anger. It seems almost necessary that we often join with others in declaring a likeness that serves to differentiate ourselves and perhaps that we dislike those that are not like us. This distinction of dislike can easily grow in intensity to extreme dislike or hate and so can be accompanied by hate and anger. At times, and for some, we require a constant infusion of hate and anger to both keep our group cohesion and our individuality intact.

Just as we use the selfobjects in our group of like-minded individuals to maintain ourselves, so too may we use the "others" to serve as enemies who also keep us intact. The rage and hate that we may employ against those whom we dislike can easily become a necessary component of our own existence. Without the rage we may encounter self-depletion and potential self-fragmentation. For some the rage itself can be both debilitating and a step toward fragmentation, and so is to be avoided. When one speaks of rage as addicting we thereby recognize its potential to insinuate itself into a necessary component of self-identity and self-esteem. It is essential that it be evaluated without any prior prejudicial position. Some people need to be angry to feel alive.

It may be too simplistic to claim empathy as an answer to rage inasmuch as the rageful person is never interested in understanding the "other." Understanding another tends to diminish differences, and these differences are vital in sustaining anger. If we add a moral dimension to understanding then all hope for hate may be lost. One can employ empathy and understanding in order to further distinguish oneself from the other, just as the canny salesman can "understand" the potential customer just enough to take advantage of him (or her) but never to the point of becoming as one with him or her. Perhaps we should consider empathy as a two-step process. The German dive bombers who used sirens to terrify the people who were to be bombed were empathic only to the point of knowing what it was like to be frightened but not to the point of a total identification with their victims.

When a psychoanalyst aims to understand an angry, rageful patient, he or she has to join in the anger without a full participation with the anger, but one may well run the risk of not fully understanding the patient until such a full experience can be realized. This is so because anger and rage are such powerful signals of difference. The start of the process of understanding is the recognition of the patient's need to be outraged and belligerent without the analyst considering it as a symptom to be rid of. The diminution of this need can be achieved only with the employment of selfobjects which bolster the self and enhance self-esteem without the insistence on being different from the "other," who is often cast as the enemy. The most common transference configuration is that of allegiance to the analyst as a member of the group and/or anger at the analyst for being outside of the group. As the rage diminishes the emergence of depression (which is usually that of an empty depression) occurs. Soon the search for selfobjects, which raise self-esteem without deprecating others, appears.

The familiar advice to "love your enemy" invites us to examine the entrance of morality into the problem of rage. It is commonplace to consider hate and rage as negative affects which can be alleviated or removed by the positive maneuvers of understanding and caring. Of course, moral concerns do run counter to hate, but can also be counterproductive to completely understanding the need to hate. As contradictory as it may seem we need to join the hate on the road to its elimination. The first step on the road is that of seeing the positive side of outrage.

Chapter 12

Carving out a place for psychoanalysis

In a far-ranging book dealing with art, psychology and neuroscience Eric Kandel reveals a fundamental flaw of his own efforts, along with those of the new champions of neuroscience, to join psychoanalytic thinking with an emphasis on the brain and biology (Kandel, 2012). Kandel makes the simple conflation of "not conscious," which he attributes to the vast majority of the brain's activity, with "unconscious," which he sometimes rather loosely relates to the psychoanalytic conception of the repressed unconscious or the dynamic unconscious. Kandel writes as if he is familiar with psychoanalytic thinking, but he does not seem to recognize that the unconscious of psychoanalysis does not refer at all to the many brain processes that ever escape our awareness. Kandel's infatuation with neuroscience leads him to a corresponding disappointment in the failure of psychoanalysis to pursue what he so champions: empirical studies.

This flaw that is so clearly presented by Kandel is essentially a confusion of vocabulary and is one primary reason for the growing separation of psychiatry and psychoanalysis. For one, it is rather routine that a given bit of behavior can hopefully, over time, be traced to a particular area of the brain and so is said to be "caused" by that regional activation. We readily say that reaching out my arm for my cup of coffee is caused by the motor neurons in the motor area of the brain. There is usually no cause for argument there. However, that movement is also clearly "caused" by my wish to drink some coffee. We somehow manage to have two causes, and we may divide them into proximal and distal, or primary and secondary, or might even claim one is "correlated with" and one is "caused by." We may attempt to determine the real cause by introducing a time line – i.e., I wanted the coffee and so activated that particular motor area – but Libet (1985) has shown us that a "readiness potential" or a blip in the electrical record of the brain occurs *before* the conscious wish to activate a movement. But, of course, we may have unconsciously wanted it before we became conscious of our desire. However, surely that very wish is a product of brain activity as well, since the mind must result from the operations of the brain. However, we must beware of a reductionist fallacy, inasmuch as reducing it all to neuronal activity will reveal no more than reducing Shakespearean sonnets to the mere letters of the words. Essentially, these are two quite different universes of discourse; i.e., our words and sentences in the one are simply not exchangeable. The brain and the mind are both interdependent as well as presently independent and incapable of unity. Sad, but true. Any effort to reduce the one to the other loses the essence of each.

As Paul Ricoeur said of the language of the brain versus that of the mind, "these discourses represent heterogeneous perspectives which is to say that they cannot be reduced to one another or derived from one another. I therefore combat that sort of semantic amalgamation" (Changeux & Ricoeur, 2000). The same may be true, in a starker manner, of an attempt of the above-mentioned effort to reduce a Shakespearian sonnet to its letters. We cannot derive the beauty or meaning of a sonnet from a diligent study of the frequency or arrangement of the letters or even of the words. A careful study of the pixels that activate a television screen will not tell us the score of the basketball game that is being televised. We employ different universes of discourse that cannot be reduced or derived from one another. The struggle over the persistent search for the determination of just when and how the neurons of the brain give rise to consciousness is not very much different than that of determining when the sounds we make manage to form words which we understand, nor when the letters of our alphabet magically form a meaningful sentence. This is much like the advice given to a lost traveler: "you can't get there from here." No study of the brain will reveal the mind. The brain does not think or plan or have meaning. Those are provinces of the mind. Indeed, when efforts are made to equate the brain with the mind, the brain seems to become a person who makes decisions, sends messages and evaluates the results of its efforts. We see the brain as a computer, which has no little person inside but merely follows the code of its software.

Psychiatry has ever been involved in a search for primary causes, and this effort has traveled the long road from demons to bad upbringing to brain aberrations. When Sigmund Freud presented psychoanalysis to the world, psychiatry embraced it and offered it the principle position as an explanation of mental illness. Up until then, psychiatry had devoted itself to categorizing and classifying mental illness with but a few resources for treatment. With psychoanalysis aboard, it was able to both treat as well as to explain mental illness and could even refine some of its categories. The love affair with psychoanalysis dominated the psychiatric field until the introduction of a variety of psychotropic drugs, and soon thereafter the effectiveness of the new treatments shifted attention to the brain and those areas of the brain that seemed central to these treatments, and so it was but natural that that might better serve to explain the causes of mental illness.

Much as a newfound lover upsets a somewhat successful marriage, the advent of psychotropic drugs and the subsequent investigation of the brain brought about the inevitable trial separation and subsequent impending divorce of psychiatry and psychoanalysis. The problem that remained had to do with custody, and, rather than giving it all to the one or the other, mental illness became the province of both psychiatry and its newfound marriage to the brain, along with psychoanalysis and *its* own companion and partner of psychodynamic psychotherapy. Once again the parallels to love and marriage are familiar, as so many of those involved in the marriage from friends and relatives to realtors to clergy are ever hopeful of reconciliation and reunion. Not surprisingly, an equal number of interested parties insist that these fields were incompatible from the start, and they need both go their separate ways. In the subsequent distribution of the marital assets most of the material possessions seemed to go to psychiatry and the neurosciences, whose supporters in the form of the insurance companies and the pharmaceutical industry gave it the bulk of the assets. Psychoanalysis is now in the unenviable position of finding both a place to live as well as a function to perform. It needs to go its own way, and in the corniest of phrases, it needs to find itself.

Mental illness therefore seems to have been placed into shared custody and thus can only be described as torn between living with psychiatry and neuroscience, with their promises of quick relief and rapid but often partial satisfaction, or taking up the familiar and sometimes pleasant residence with psychoanalysis and psychotherapy, with both the expense and uncertain future. In a sense it is unfair to ask mental illness to decide inasmuch as the more powerful forces of decision lie outside of its domain. Its fate is decided by others.

Psychoanalysis has long sort of hitchhiked on the diagnostic classification of psychiatry, both because Sigmund Freud did so and because there was no pressing need to develop its own categories. It too made its diagnoses because of its continuing relationship to psychiatry along with its dependence on insurance reimbursement. However, left to its own devices, it might well be found to dispense with diagnostic categories in favor of vague concepts such as analyzability or, as its cousin of psychodynamic psychotherapy might in turn say, it is directed toward the even vaguer concept of treatability.

If psychoanalysis and psychotherapy were to admit that they really devote their efforts to the above-named categories, by necessity they would also have to recognize that they have vacated a host of psychiatric diagnoses ranging from alcoholism to the myriad forms of substance abuse to the large group of psychotic disorders. These vacated areas have been quickly filled by a variety of efforts ranging from modified analysis, which may at times claim a certain authenticity, to modified psychotherapy, which may disregard diagnosis entirely. These modifications often emphasize the "relationship" or the personal qualities of the therapist, but for the most part diagnosis seems to take a secondary role in the decision of applicability. The vacated areas became arenas of conflict over ownership.

Psychoanalysis has been forced to reckon with the fact that it is designed to allow patients to better understand themselves by an increased awareness of their unconscious processes, and this may or may not (but often does) lead to increased well-being and symptom relief. It does not nor could it ever claim effectiveness in the areas that psychiatry wishes to dominate, although it often competes with psychiatry in some of those areas. Indeed, if psychoanalysis was forced to delineate where and when it had a role it might well decide on something like "ordinary unhappiness" with a sigh of relief that it never need be more specific. It may further claim that it does not do diagnostic assessments as psychiatry does, but rather develops a diagnosis over time and with the aid of the developing transference. Depending on the particular school of psychoanalysis that the practitioner belonged to, the diagnosis may range from "narcissistic personality disorder" to something borrowed from psychiatry, such as "borderline personality," which has become a barometer of the likeability of the patient. Perhaps diagnosis is yet another asset of the marriage between psychoanalysis and psychiatry that should be relinquished to psychiatry while analysis and therapy begin an independent existence that was not planned for.

The independence of psychoanalysis – its recognition

A recent review of a book on psychoanalysis in America reached the conclusion that psychoanalysis, having achieved great prominence after 1945, had subsequently collapsed. The reviewer defended this point by noting that the American Psychoanalytic Association reported a 50% decline over a 30-year period in the number of applicants for training, with what they called "an even more precipitous decline in applications from psychiatrists." It also noted that the profession is aging rapidly. The International Psychoanalytical Association announced that 70% of its membership was aged between 50 and 70: 50% were older than 60, and as many as 20% of training analysts were over the age of 70 (Scull, 2012). Analysts no longer head university departments of psychiatry, and psychologists and social workers have become the primary providers of counseling, often relying on cognitive behavioral therapy as the intervention of choice.

Whether or not one wishes to declare the death of psychoanalysis, it does seem advisable and even necessary to recognize that it has little or nothing to do with psychiatry, has a questionable and arguable relationship to mental illness and is presently in a desperate need to re-establish and re-define itself. It must begin by staking out an independent claim of existence.

It would seem likely that the present state of psychoanalysis will continue to decline or deteriorate in its relationship to psychiatry, its source of new practitioners or those at all interested in the field and its pertinence to the world. Once this status or position is recognized it can attend to its new and independent place. The denial of these facts will impede and unfortunately even prevent a new place for psychoanalysis. This denial can be seen in the embrace of neuroscience by the adherents of a field called neuropsychoanalysis, by one or another specialized schools of psychoanalysis which may lay claim to a universal applicability as seen in followers of the various personalities in psychoanalysis and finally by those who insist that nothing need be changed. This last position can be seen at work in the endless and futile debates about certification in psychoanalysis, along with the struggle over the once important, but probably no longer relevant, position of training analysts. These debates at heart may be heated but are essentially empty, because they fail to have any impact outside of the increasingly small groups that argue about them. Only when psychoanalysis awakens to its new, independent and often lonely existence can it begin to change.

Clinical illustration

The ongoing struggle in the study of mental health disorders is reflected in the lack of precision in the proper positioning of the brain, the mind and the self. This is somewhat parallel to the story of a composite patient, whom we shall call Anne, and her journey in seeking proper treatment. Anne came to psychoanalysis after a number of psychotherapeutic experiences. She felt that she had profited from all of her therapy in that it was helpful to see someone regularly and to have someone to talk to. There were no particular insights that she recalled from her therapy, although the therapist did make a number of comments about separation anxiety. Indeed, Anne did not seem to have a clear-cut problem that would easily fit in the DSM-5, save that of a chronic feeling of dissatisfaction and unreality. Anne felt that her psychoanalysis was different in that it enabled her to construct a time line of her life, a historical narrative or perhaps, better put, an autobiography. She recognized that this autobiography was not written entirely by her and that it included memories that she could never have recovered without the help of her analyst, who was more properly seen as a co-author. Seeing herself over time gave Anne an understanding of her life that enabled her to follow the varied forms of Anne that emerged over the years.

One striking event in Anne's life was the death of someone quite close to her when she was a child. She had spoken of this death during psychotherapy and much had been made of her inability to fully mourn this person. One may see the parallel to the present uncertainty in normal grief from pathological mourning with the utilization of a time period. Anne, for one, had gotten over her grief in one sense, perhaps even with medication, but not in another, in that only in psychoanalysis was she able to fully grasp what this person and the loss had really meant to her. The re-experience of loss in the analysis gave a clarity to the event that was unavailable in the many other efforts that Anne pursued to deal with it. There can be no doubt that Anne's brain changed at the time of the loss and subsequently. However, how she thought about the loss – i.e., what it meant to her – and how she became a different person with the loss and with this new understanding were other ways to conceptualize the impact on Anne of her pre- and post-analytic selves.

Most people do, or can with some effort, construct some form of a biographical account of themselves which may highlight standout events such as graduation, marriage and divorce, status, etc. Painful issues play a role in these histories, but not everyone can successfully process the varied sources of pain, and psychological symptoms of one kind or another also become historical moments. Psychotherapy based on psychoanalytic principles may alleviate such symptoms, just as many forms of treatment may be successful in this pursuit. Indeed, many individual historical narratives may be free or almost free of problematic moments. However, the full autobiography is riddled with missing parts that only a psychoanalysis can bring to light. One ends with the conclusion that not everyone needs an analysis, but perhaps everyone can use an analysis.

From one viewpoint that utilizes one language and one set of theoretical propositions, Anne's brain reacted to her loss with certain neurochemical alterations, which in turn could be modified or perhaps masked by medication. From another point of view that utilizes a different language and is based upon different theoretical presumptions, Anne's feelings were altered in her psychotherapy and the meanings assigned to the events of her life were altered. And yet from still another set of theoretical ideas, Anne's conceptualization of her life's history and Anne's perception of her person took on a new meaning in her psychoanalysis. Each of these ways of intercession upon Anne's psychology had certain merits and certain failings, but each was different and meaningful. However, their differences are not to define them as exclusive inasmuch as everything mental is in the brain, all meaning is within the mind and the self or person encompasses one's entire history. They use a language and set of concepts designed to evoke a particular group of outlooks and insights with no need for destructive competition. They cover a similar area of concern but approach it with a different set of tools and with different results.

The independence of psychoanalysis – its opportunity

If one believes that psychoanalysis is a body of knowledge, a therapeutic technique and an instrument of research and so has much to offer, then it remains as a task ahead to find and develop a proper path for it to pursue. Psychoanalysis surely is able to better delineate just who might profit from it and so might develop its own diagnostic categories. These categories should be based on the data derived from the analytic process rather than that of overt behavior or personal descriptions of discontent.

Psychoanalysis is often cited as a useful research tool in many disciplines outside of psychiatry. However, even within psychiatry we can see how it can and should make a contribution. So many research efforts have been influenced, or some say plagued, by the placebo effect. Despite the fact that this effect does indeed become reflected in brain areas of hyperactivity, there remains a need for determining the psychologic factors responsible for its presence. For the most part the placebo effect has been seen as an interference in assessing the efficacy of certain medications rather than as an opportunity to better understand the meaning of the effect. As noted by Hedges and Burchfield (2005),

> The placebo response appears to occur across a variety of clinical conditions and seems to depend upon expectation to elicit its effects. . . . Far from an embarrassing finding to be swept away with hushed discussions of effect sites, power and statistical significance, the placebo effect may be integral to a comprehensive understanding of disorder and disease.

It should be examined, not as an interference, but for what it means in each individual. Once again, psychoanalysis might profit from freeing itself from psychiatry.

All in all, in recruitment and training, in diagnosis and treatment, in the very future of its existence, psychoanalysis is in need of its own emancipation proclamation.

Summary

As much as psychoanalysis has benefited from its relationship with psychiatry, it has also suffered from it, and as psychiatry becomes more enamored of, and dependent on, its own turn to and involvement with neuroscience, analysis must decide whether to join in this pursuit or to free itself from the new and now central repositioning of psychiatry. To be free of psychiatry means to be free of its categories of diagnosis, its recruitment of physicians with their connections to insurance and pharmaceutical companies and especially of its language and its preoccupation with the discourse of the brain. A free and independent psychoanalysis may well appeal to younger university students, as well as direct attention to many neglected areas for future research.

The task ahead for an independent psychoanalysis is to carefully determine, with more precision than it presently possesses, just who can and should be analyzed. Analysis should be positively indicated rather than being done if nothing else works or else being tried on everyone who agrees to participate. A better application of the technique should allow for psychoanalysis to have its own diagnostic categories without modifying them to satisfy problems of reimbursement.

Psychoanalysis is the discipline that studies the role and meaning of the contributions of unconscious mentation to our lives. It is neither the study of the brain nor of the many other discourses that may present themselves for inquiry. Whether or not one agrees with this position, history will decide its fate.

Chapter 13

The future
Epilogue

The advice that may be offered to the medical student who is interested in psychiatry, and therefore in the workings of the brain, is necessarily different from that offered to the college student who may be interested in the mind or the concept of agency – i.e., the self. Each of them may have occasion to learn something of what the other is doing, but over time the overlap may well be seen to diminish. The popular items in the news have mainly to do with the determination of specific areas of the brain being devoted to specific sorts of mental activity, while the question or questions of trying to ascertain the meaning or meanings of why a person did or felt something usually becomes relegated back to the study of the brain. The mind takes a back seat, while the self hardly gets noticed.

The issue that brings these two students together, if it is ever to be acknowledged, is one of treatment. The happy discovery that psychotherapy brings about certain alterations in the brain (Gabbard et al., 2012) allows for a "meeting of the minds" which offers grounds for a possible collaboration. However, while the neuroscientist may cheer the changes in brain chemistry that result from "talk therapy," it is a rare therapist who makes the claim that transmagnetic stimulation allowed a patient to better understand himself or herself. Feeling better is an endpoint in treatment, while self-understanding is almost seen as an indulgence. Drugs do not and should not lead to insight. Indeed, insight itself is often seen as a byproduct of "talk therapy." What good is insight if one is still unhappy? Alas, the efforts of psychoanalysis are of no use without the guarantee of relief from suffering, and a tighter fit between understanding and feeling better is a necessary requirement if psychoanalysis is to endure. Perhaps here is where the self can make a claim for recognition as the agent of self-empathy; i.e., to the degree that I am aware of and in touch with my history my self-understanding will put me in charge of my future. No drug can or should do that, and not everyone can or should make that choice.

Of course, prediction is a risky activity, and all sorts of scenarios may play out in the future of psychoanalysis and brain science. However, right now it looks as if they will grow further apart, and psychodynamic psychotherapy will be but an occasional visitor to psychiatry, while the study of the brain will continue to be a nagging interloper in the many schools devoted to psychoanalysis.

Psychoanalysis per se shall probably become a historical oddity in the study of psychiatry. Inasmuch as the many voices of psychoanalysis hardly have much to say to one another these days, it seems unlikely that they will share a commonality with neuroscience. Perhaps a new language will emerge that succeeds in joining these isolated arenas of discourse, both within psychoanalysis as well as between neuroscience and psychoanalysis. We are presently too involved in the choice of champions, and so true partnerships appear to be quite away in the future. But one never knows.

Psychoanalysis would be at its best if it could see this moment as one of opportunity. Perhaps the alliance with psychiatry and its eagerness to embrace neuroscience has been an unfortunate form of bondage that has kept analysis from its own growth and development. Perhaps we have too long ignored a potential that now beckons to us if only we would listen.

Bibliography

Abend, S. M., Porder, M. S., & Wallack, M. A. (1983). *Borderline Patients. Psychoanalytic Perspectives.* New York: International Universities Press.
Adler, G. (1977). *Borderline Psychopathology and Its Treatment.* New York: Jason Aronson.
Adler, G. (1988). How useful is the borderline concept? *Psychoanalytic Inquiry, 8*: 353–372.
Adler, G., & Buie, D. (1979). Aloneness and borderline psychopathology: The possible relevance of child developmental issues. *International Journal of Psychoanalysis, 60*: 83–96.
Agosta, L. (2009). *Empathy in the Context of Philosophy.* New York: Palgrave Macmillan.
Alexander, F. (1961). *The Scope of Psychoanalysis, 1921–1961: Selected Papers.* New York: Basic Books.
American Psychiatric Association. (2000). *Diagnostic and Statistical Manual of Mental Disorders* (4th ed., text revision). Washington, DC: Author.
Angell, M. (2011, July 14). The illusions of psychiatry. *The New York Review of Books, LVIII*(12): 20–24.
Aragno, A. (2008). The language of empathy: An analysis of its constitution, development, and role in psychoanalytic listening. *Journal of the American Psychoanalytic Association, 56*: 713–740.
Arlow, J. (1961). Silence and the theory of technique. *Journal of the American Psychoanalytic Association, 9*: 44–55.
Bae, S., Hashimoto, H., Karlson, E., Liang, M., & Daltroy, L. (2001). Variable effects of social support by race, economic status, and disease activity in systemic lupus erythematosus. *J. Rheumatology, 28*(6): 1245–1251.
Baron-Cohen, S. (1995). *Mindblindness: An Essay on Autism and Theory of Mind.* Cambridge, MA: MIT Press.
Baron-Cohen, S. (2011). *The Science of Evil: On Empathy and the Origins of Cruelty.* New York: Basic Books.
Basch, M. F. (1983). Empathic understanding; A review of the concept and some theoretical considerations. *Journal of the American Psychoanalytic Association, 31*(1): 101–126.
Batson, C. (2009). These things called empathy: Eight related but distinct phenomena. In J. Decety & W. Ickes (Eds.), *The Social Neuroscience of Empathy* (pp. 3–15). Cambridge, MA: MIT Press.
Bergmann, M. (1993). Reflections on the history of psychoanalysis. *Journal of the American Psychoanalytic Association, 41*: 924–955.
Bernstein, R. (2002). Hermeneutics, critical theory, and deconstruction. In R. J. Dostal (Ed.), *The Cambridge Companion to Gadamer* (pp. 267–282). Cambridge: Cambridge University Press.

Bischol-Kohler, D. (1988). On the association of empathy and the ability to recognize oneself in a mirror. *Schweizersiche Zeitschrift fur Psychologie, 47*: 147–259.

Brandchaft, B., & Stolorow, R. (1985). Reply to Melvin Sabshin on the question of nonmedical candidates. *Journal of the American Psychoanalytic Association, 33*(4): 970–971.

Brandom, R. B. (2008). *Between Saying and Doing: Towards an Analytic Pragmatism.* Oxford: Oxford University Press.

Brent, D.A., Emslie, G., Clarke, G., Asarnow, J., Spirito, A., Ritz, I., . . . Keller, M. B. (2009). Predictors of spontaneous and systematically assessed suicidal adverse events in the treatment of SSRI-resistant depression in adolescents (TORDIA) study. *American Journal of Psychiatry, 166*(4): 418–426.

Brockbank, R. (1970). On the analyst's silence in psychoanalysis. A synthesis of intrapsychic conflict and interpersonal manifestations. *International Journal of Psychoanalysis, 51*: 457–464.

Brooks, D. (2009, October 13). The young and the neuro. *The New York Times.*

Cacioppo, J., Hawkley, L. Crawford, L., Ernst, J., Burleson, M., Kowalewski, R., . . . Berntson, G. (2002). Loneliness and health: Potential mechanisms. *Psychosomatic Medicine, 64*: 407–417.

Caligor, E., Stern, B., Hamilton, M., MacCormack, V., Winiger, L., Sneed, J., & Royse, S. (2004). Why we recommend analytic treatment for some patients and not for others. *Journal of the American Psychoanalytic Association, 57*(5): 677–694.

Carruthers, P. (2009). How we know our own minds: The relationship between mind reading and metacognition. *Behavioral and Brain Science, 32*: 121–182.

Caston, J. (2011). Agency as a psychoanalytic idea. *Journal of the American Psychoanalytic Association, 59*(5): 907–938.

Changeux, J. P., & Ricoeur, P. (2000). *What Makes Us Think?* Princeton: Princeton University Press.

Churchland, P. (1989). *Neurophilosophy: Toward a Unified Science of the Mind-Brain.* Cambridge, MA: MIT Press.

Clark, A. (2013). Predictive brains, situated agents, and the future of cognitive science. *Behavioral and Brain Sciences, 36*: 181–253.

Coates, S. (1977). Is it time to jettison the concept of developmental line? *Gender and Psychoanalysis, 2*: 35–53.

Cohen, J. (2010). *Almost Chimpanzee: Searching for What Makes Us Human, in Rainforests, Labs, Sanctuaries, and Zoos.* New York: Times Books.

Corbett, K. (2011). Gender regulation. *Psychoanalytic Quarterly, LXXX*(2): 441–459.

Coyne, J. (2009). *Why Evolution Is True.* New York: Viking.

Crick, F. (1994). *The Astonishing Hypotheses: The Scientific Search for the Soul.* London: Simon and Schuster.

Decity, J., & Meyer, M. (2008), From emotion resonance to empathic understanding: A social developmental neuroscience account. *Development and Psychopathology, 20*:1053-1080.

Denberg, T.D., Melhado, T. V., Coombes, J. L., Beaty, B., Berman, K., Byers, T., . . . Ahnen, D. (2005). Predictors of non-adherence to screening colonoscopy. *Journal of General. Internal Medicine, 20*: 989–995.

Dennett, D. (2013a). Expecting ourselves to expect. The Bayesian brain as a projector. *Behavioral and Brain Sciences, 36*(3): 209–210.

Dennett, D. (2013b). *Intuition Pumps and Other Tools for Thinking.* New York: W.W. Norton & Co.

de Waal, F. (2009). *The Age of Empathy: Nature's Lessons for a Kinder Society*. New York: Harmony.

Dewey, J. (1939). *Intelligence in the Modern World*. New York: Modern Library.

Dilip, J., Lieberman, J., Scully, J., & Kupfer, D. (2012). DSM crosses the finish line. *Psychiatric News, 47*(24): 4–23.

Dostal, R. (2002). *The Cambridge Companion to Gadamer*. Cambridge: Cambridge University Press.

Dreyfus, H. (1991). Heidegger's hermeneutic realism in the interpretive turn. In D. Hiley, J. F. Bohman, & R. Shusterman (Eds.), *The Interpretive Turn: Philosophy, Science, Culture* (pp. 25–41). Ithaca: Cornell University Press.

Eco, U. (2013). *Inventing the Enemy. Essays*. New York: Mariner Books.

Ekman, P., Levenson, R. W., & Friesen, W. V. (1983). Autonomic nervous system activity distinguishes among emotions. *Science, 221*: 1208–1210.

Elqayam, S., & Evans, J. St. B. T. (2011). Subtracting "ought" from "is": Descriptivism versus normativism in the study of human thinking. *Behavioral and Brain Sciences, 34*(5): 233–275.

Fearn, N. (2005). *The Latest Answers to the Oldest Questions*. New York: Grove Press.

Ferenczi, S. (1913). *Sex in Psychoanalysis: The Selected Papers of Sándo Ferenczi*, vol. 1. New York: Basic Books [1950], pp. 213–239.

Fonagy, P., Gergely, G., Jurist, E., & Turget, M. (2002). *Affect Regulation, Mentalization, and The Development of the Self.* New York: Other Press.

Fonagy, P., & Turget, M. (2006). The mentalization-focused approach to self pathology. *Journal of Personality Disorders, 20*: 544–576.

Freud, A. (1946). *The Ego and the Mechanisms of Defense*. New York: International Universities Press.

Freud, A. (1965). *Normality and Pathology in Childhood*. New York: International Universities Press.

Freud, S. (1900). The Interpretation of Dreams. *Standard Edition, IV*.

Freud, S. (1905a). Dora: An analysis of a case of hysteria. *Standard Edition, VII*: 3–122.

Freud, S. (1905b). Three essays on the theory of sexuality. *Standard Edition, VII*: 125–231.

Freud, S. (1920–1922). Group psychology and the analysis of the ego. *Standard Edition, XVIII*: 101.

Freud, S. (1926a). The question of lay analysis. *Standard Edition, XX*: 179–251.

Freud, S. (1926b). *Standard Edition, XXI*: 119.

Freud, S. (1930). Civilization and its discontents. *Standard Edition, XVI*: 59–145.

Freud, S. (1957). The taboo of virginity. *Standard Edition, XI*: 193–208. (Original work published 1910)

Freud, S. (1964). Moses and monotheism. *Standard Edition, XXIII*: 41. (Original work published 1940)

Frith, C., & Wolpert, D. (Eds.). (2004). *The Neuroscience of Social Interaction: Decoding, Imitating and Influencing the Actions of Others*. Oxford: Oxford University Press.

Gabbard, G., Litowitz, B., & Williams, P. (2012). *Textbook of Psychoanalysis*. Washington, DC: American Psychiatric Publishing.

Gallese, V. (2008). Empathy, embodied simulation and the brain. *Journal of the American Psychoanalytic Association, 56*: 769–781.

Gaynes, D., Warden, D., Trivedi, M., Wisniewski, S., Fara, M., & Rush, A. (2009). What did STAR*D teach us? Results from a large-scale practical clinical trial for patients with depression. *Psychiatric Services, 60*(11): 1439–1445.

Gedo, J., & Goldberg, A. (1973). *Models of the Mind*. Chicago: University of Chicago Press.
Glover, E. (1950). Functional aspects of the mental apparatus. In *On the Early Development of The Mind* (pp. 369–378). New York: International Universities Press.
Goldberg, A. (1975). A fresh look at perverse behavior. *Int. J. Psych., 56*(3): 335–342.
Goldberg, A. (1999). *Being of Two Minds*. Hillsdale, NJ: The Analytic Press.
Goldberg, A. (2001a). Depathologizing homosexuality. *Journal of the American Psychoanalytic Association, 49*: 1109–1114.
Goldberg, A. (2001b). Me and Max. A misalliance of goals. *Psychoanalytic Quarterly, 70*: 117–130.
Goldberg, A. (2002). American psychoanalysis and American pragmatism. *Psychoanalytic Quarterly, 71*(1): 235–250.
Goldberg, A. (2004). *Misunderstanding Freud*. New York: Other Press.
Goldberg, A. (2007). *Moral Stealth*. Chicago: University of Chicago Press.
Goldberg, A. (2011a). *The Analysis of Failure*. New York: Routledge.
Goldberg, A. (2011b). The enduring presence of Heinz Kohut: Empathy and its vicissitudes. *Journal of the American Psychoanalytic Association, 59*: 289–312.
Goldner, V. (2011). Trans: Gender in free fall. *Psychoanalytic Dialogues, 21*(2): 159–171.
Greenson, R. (1962). On enthusiasm. *Journal of the American Psychoanalytic Association, 10*: 3–21.
Gross, C. (2013). Between brain and imaging hype. *Science, 340*(6): 1526.
Hagoort, P., & Levelt, W. (2009). The speaking brain. *Science, 326*(5951): 372–373.
Hales, R. (2013). DSM-5 self exam. *Psychiatric News*, p. 14.
Hartmann, H. (1939). *Ego Psychology and the Problem of Adaptation*. New York: International Universities Press.
Hatfield, E., Rapson, R., & Le, Y.-C. (2009). Emotional contagion and empathy. In J. Decety & W. Ickes, *The Social Neuroscience of Empathy* (pp. 19–30). Cambridge, MA: MIT Press.
Hawkley, L. G., & Cacioppo, J. T. (2003). Loneliness and pathways to disease. *Brain, Behavior, and Immunity, 17*(Supplement 1): 98–105.
Hedges, D., & Burchfield, C. (2005). The placebo effect and its implications. *The Journal of Mind and Behavior, 26*(1): 161–180.
Heidegger, M. (1927). *Being and Time* (J. Stambough, Trans.). Albany, NY: State University of New York Press [1946].
Heidegger, M. (1994). *Basic Questions of Philosophy* (R. Rojcewicz & A. Schwer, Trans.). Bloomington, IN: Indiana University Press.
Herve, H., & Yuelle, J. (2007). *The Psychopath. Theory, Research and Practice*. Mahwah, NJ: Lawrence Erlbaum.
Hoffman, M. L. (2000). *Empathy and Moral Development*. Cambridge: Cambridge University Press.
Holzman, P. (1985). Psychoanalysis: Is the therapy destroying the science? *Journal of the American Psychoanalytic Association, 33*(4): 725–770.
Horikawa, T., Tamaki, M., Miyawaki, Y., & Kamitani, Y. (2013). Neural decoding of visual imagery during sleep. *Science, 340*(6132): 639–642.
Hume, D. (1742–1754). Essay X: Of superstition and enthusiasm. In *Essays: Moral, Political and Literary* (pp. 74–80).
Hurley, S. (2008). The shared circuits model. How control, mirroring, and simulation can enable imitation, deliberation, and mind reading. *Behavior and Brain Science, 31*(1): 1–38.

Inglis, F. (2009). *History Man. The Life of R. G. Collingwood*. Princeton, NJ: Princeton University Press.

Ingram, D. (1985). *Hermeneutics and Truth in Hermeneutics and Praxis* (R. Hollinger, Ed.). Notre Dame, IN: University of Notre Dame Press.

Isay, R. (1989). *Being Homosexual*. New York: Ferrar, Straus & Giroux.

Jacobs, D. (2011). Is the DSM's formulation of mental disorder a technical-scientific term? *The Journal of Mind and Behavior, 32*(1): 63–80.

Kandel, E. (2012). *The Age of Insight. The Quest to Understand the Unconscious in Art, Mind and Brain From Vienna 1900 to the Present*. New York: Random House.

Kantrowitz, J. (1996). *The Patient's Impact on the Analyst*. Hillsdale, NJ: The Analytic Press.

Karen, R. (1990). Becoming attached. *The Atlantic Monthly, 265*(2): 35–76.

Kernberg, O. (1976). *Object Relations Theory and Clinical Psychoanalysis*. New York: Jason Aronson.

Kernberg, O. (1995). *Love Relations: Normality and Pathology*. New Haven: Yale University Press.

Khamsi, R. (2013, June). Brain scans could become EKG for mental disorders. *Time*. Available at http://healthland.time.com/2013/06/28/brain-scans-could-become-ekgs-for-mental-disorders/

Kohut, H. (1971). *The Analysis of the Self*. New York: International Universities Press.

Kohut, H. (1984). *How Does Analysis Cure?* Chicago: University of Chicago Press.

Kuhn, T. (1970). *The Structure of Scientific Revolutions*, 2nd ed. Chicago: University of Chicago Press.

Kuhn, T. (1991). The natural and the human sciences. In D. Hiley, J. Bohman, & R. Shusterman (Eds.), *The Interpretive Turn: Philosophy, Science, Culture* (pp. 17–24). Ithaca, NY: Cornell University Press.

Kurzweil, R. (2012). *How to Create a Mind: The Secret of Human Thought Revealed*. New York: Penguin.

Lane, C. (2007). *Shyness: How Normal Behavior Became a Sickness*. New Haven, CT: Yale University Press.

Leigner, E. (2003). The silent patient. *Modern Psychoanalysis, 28*: 69–83.

Lewis, M. (1993). Commentary. *Human Development, 36*: 363–367.

Libet, B. (1985). Unconscious cerebral initiative and the role of conscious will in voluntary action. *Behavioral and Brain Science, 8*: 529–566.

Lindquist, K., Wager, T. D., Kober, H., Bliss-Moresa, E., & Barret, L. (2012). The brain basis of emotion: A meta-analytic review. *Behavioral and Brain Sciences, 35*: 121–1433).

Lipps, T. (1903). Einfuhling, inner Nachahmung, und orges-empfindungen. *Achiv fur Gesamte Psychologic, 1*: 185–204.

Luzzi, J. (2013, April 21). This Could be "Heaven" or This Could be "Hell." *New York Times Book Review*, 13.

MacCulloch, D. (2013). *Silence: A Christian History*. New York: Viking.

MacIntyre, A. (1985). *Where Justice? Which Rationality?* Notre Dame, IN: University of Notre Dame Press.

Maturana, H. R., & Varela, F. J. (1992). *The Tree of Knowledge: The Biological Roots of Human Understanding*. Boston: Shambhala, 255.

McWilliams, N. (1994). *Psychoanalytic Diagnosis. Understanding Personality Structure in Clinical Process*. New York: The Guilford Press.

Metzinger, T. (2009). *The Ego Tunnel*. New York: Basic Books.
Miller, G. (2013). The promise and perils of Oxytocin. *Science, 339*: 267–269.
Mitchell, S. (1991). Wishes, needs, and interpersonal negotiations. *Psychoanalytic Inquiry, 11*: 147–170.
Mlodinov, L. (2012). *Subliminal. How Your Unconscious Mind Rules Your Behavior*. New York: Pantheon Books, 93–96.
Moran, M. (2013). Eating, sleep disorder criteria revised in DSM-5. *Psychiatric News*, p. 14.
Nagel, T. (2012). *Mind and Cosmos: Why the Materialist Neo-Darwinian Conception of Nature is Almost Certainly False*. Oxford: Oxford University Press.
Nöe, A. (2004). *Action in Perception*. Cambridge, MA: MIT Press.
Nöe, A. (2009). *Out of Our Heads: Why You Are Not Your Brain and Other Lessons From the Biology of Consciousness*. New York: Hill and Wang.
Offer, D., & Sabshin, M. (Eds.). (1984). *Normality and the Life Cycle*. New York: Basic Books.
Orwell, G. (1993). *1984*. New York: Plume.
Panksepp, J. (1998). *Affective Neuroscience: The Foundation of Human and Animal Behavior*. Oxford: Oxford University Press.
Piaget, J. (1928). 1973 *The Child's Conception of the World*. Abingdon and New York: Routledge.
Popper, K. (1994). *The Myth of the Framework*. London: Routledge.
Premack, P., & Woodruff, G. (1978). Does the chimpanzee have a theory of mind? *Behavioral and Brain Sciences, 1*(4): 515–526.
Pres, R. (2013). Letters to the editor. *The New York Times*, p. A22.
Preston, S. D., & de Waal, F.B.M. (2002). Empathy: Its ultimate and proximate bases. *Behavioral and Brain Sciences, 25*: 1–72.
Quine, W. V. (1987). *Quidddities*. Cambridge, MA: The Belknop Press of Harvard University Press.
Reik, T. (1968). The psychological meaning of silence. *Psychoanalytic Review, 55*: 72–186.
Rizzolotti, G., Fadiga, L., Fogassi, L., & Gallese, V. (1999). Resonance behavior and mirror neurons. *Archives Italienne de Biologie, 137*: 85–100.
Rizzolatti, G., Fadiga, L., Gallue, V., & Fogaise, L. (1996). Premotor cortex and the recognition of mirror actions. *Cognitive Brain Research, 3*(2): 131–141.
Sacks, O. (2012). *Hallucinations*. Toronto: Alfred A. Knopf.
Schafer, R. (1959). Generative empathy in the treatment situation. *Psychoanalytic Quarterly, 28*: 342–373.
Schafer, R. (1968). *Aspects of Internalization*. New York: International Universities Press.
Schafer, R. (1997). *The Contemporary Kleinians of London*. New York: International Universities Press.
Schwaber, E. (1986). Reconstruction and perceptual experience. Further thoughts on psychoanalytic listening. *Journal of the American Psychoanalytic Association, 34*: 911–932.
Scull, A. (2012, November 23). In dollaria. *Times Literary Supplement*. Available at http://www.the-tls.co.uk/tls/public/tlssearch.do?querystring=andrew+scull&offset=125&hits=0&sortby=date&order=ASC
Shedler, J. (2010). The efficacy of psychodynamic psychotherapy. *American Psychologist, 65*(2): 98–109.
Sheldrake, R. (2012). *Science Set Free*. New York: Deepak Chopra Books.

Sohms, M., & Turnbull, O. (2011). What is neuropsychoanalysis? *Neuropsychoanalysis, 13*(2): 133–145.

Stepansky, P. (2009). *Psychoanalysis at the Margins*. New York: The Other Press.

Stott, R. (2012). *Darwin's Ghosts: The Secret History of Evolution*. New York: Spiegel & Grau.

Strawson, G. (2006). Realistic monism: Why physicalism entails panpsychism. *Journal of Consciousness Studies, 13*: 3–31.

Strawson, G., Carruthers, P., Jackson, F., Lycan, W. G., McGinn, C., Papineau, D., . . . Smart, J.J.C. (2006). *Consciousness and its Place in Nature: Does Physicalism Entail Panpsychism?* (A. Freeman, Ed.). London: Academic Press UK.

Taylor, C. (2002). Gadamer on the Human Sciences. In R. J. Dostal (Ed.), *The Cambridge Companion to Gadamer* (pp. 126–142). Cambridge: Cambridge University Press.

Thomson, J. (2008). *Normativity (The Paval Carers Lectures)*. Chicago: Open Court.

Tomkins, S. (1981). The quest for primary motive: Biography and autobiography of an idea. *Journal of Personality and Social Psychology, 41*(2): 306–329.

Tuckett, D. (2005). Does anything go? *International Journal of Psychoanalysis, 86*: 31–49.

Tuckett, D., Basile, R., Birksted-Breen, D., Böhm, T., Denis, P., Ferro, A., . . . Schubert, J. (2008). *Psychoanalysis Comparable and Incomparable. The Evolution of a Method to Describe and Compare Psychoanalytic Approaches*. London: Routledge.

Van Anders, S. M., & Watson, N. (2007). Testosterone levels in women and men who are single, in long-distance relationships, or same-city relationships. *Hormones and Behavior, 51*(2): 286–291.

Varela, F.J., Thompson, E., & Rosch, E. (1991). *The Embodied Mind: Cognitive Science and Human Experience*. Cambridge, MA: MIT Press.

Vivona, J. M. (2009). Leaping from brain to mind: A critique of mirror neuron explanations of Countertransference. *Journal of the American Psychoanalytic Association, 57*: 525–550.

Wachterhauser, B. (2002). Getting it right: Relativism, realism and truth. In R. J. Dostal (Ed.), *The Cambridge Companion To Gadamer* (pp. 52–78). Cambridge: Cambridge University Press.

Wallerstein, R. (2012, August 20). Book review: *The Analysis of Failure*. *International Journal of Psychoanalysis, 95*(4): 1076–1079.

Webster's Third International Dictionary. (1965). Springfield, MA: G & C Merriam Co.

Weston, D. (2008). *The Political Brain: The Role of Emotion in Decoding the Fate of the Nation*. New York: Public Affairs.

Wilson, E. (2012). *The Social Conquest of Earth*. New York: Liveright Pub. Corp. and W. W. Norton & Co.

Wilson, M. (2006). *Wandering Significance*. Cambridge, MA: Harvard University Press.

Wispe, L. (1987). History of the concept of empathy. In N. Eisenberg & J. Strayer (Eds.), *Empathy and Its Development* (pp. 17–37). Cambridge: Cambridge University Press.

Index

Adler, Gerald 70–4
"affect storms" 85
agent vs. host 55–6
aggression, types of 136; *see also* anger; rage
Agosta, L. 94
Alexander, Franz 25
American Psychiatric Association (APA) 55
Analysis of Failure, The (Goldberg) 56
analytic relationship 29
Angell, Marcia 55
anger 135; empathy and 86; *see also* rage
Araqno, A. 89
Arat, Daniel 55
Arlow, Jacob A. 37
autism and empathy 85

Baron-Cohen, S. 44–6, 82
Basch, Michael Franz 95
Batson, C. 78
behavioral neurobiology 47
"being kept in mind" 69, 74–5; alternative theoretical model 71–2; clinical illustrations 70–1, 73–4; development and 69–70; diagnosis, psychopathology, and 70
Bergmann, Martin S. 109, 115, 118
Bible 33, 113
blank screen 36
borderline personality disorders 85
borderline psychopathology 70–2
brain: mind extending beyond the 8; self, mind, and 8; *see also* neurobiology; neuroscience
brain vs. mind debate 2, 7
Brandom, R. B. 19
Burchfield, C. 148

Carruthers, P. 82, 83
cases 20–2, 42–3, 57–8, 70–1; Anne 147–8; Charles 91–2; Dr. A. 124–5, 127; Dr. B. 125–8; Dr. F. 73, 136–8, 141; Dr. K. 133; Dr. S. 10–11; Elizabeth 92–3; Eloise 43–4; Eric 60–1; G. 31–3; Gerald 62–3; H. 102, 103; J. 102–3; Kevin 43, 44; Lisa 104–5; Marvin 57; Mike 93–4; Mr. B. 127; Mr. K. 138–9; Mr. M. 132–3; Ms. A. 127; Paul 36–7; Ronald 23, 24; Sally 59–60
Changeux, J. P. 45, 81, 144
Churchland, Paul 13, 14
cognition and empathy 89
Collingwood, R. G. 90
connectedness, development, and understanding 25
constructionist account of emotions, psychological 47
contagion, emotional 82
conviction, sense of 35
Corbett, Ken 61
corrective emotional experience 25
Crick, Francis 8, 9, 11

defenses 47–8
denial 47
Dennett, Daniel 2, 10–14
depression 131–2; anger, rage, and 135, 139, 142; case material 10–11, 23, 24, 43–4, 57, 58, 62–3, 124–5, 127; enthusiasm and 124–5, 130–2; narcissistic injury and 139; symptoms 131, 132, 134n; treatment 10–11, 18–19, 23, 41, 43–4, 48, 56, 62
developmental arrests 25, 30, 43
developmental deficits 25, 30, 33–4

de Waal, F.B.M. 77, 79
Dewey, James 46, 89
diagnosis(es): advantage of psychoanalytic 62–3; and psychopathology 70
Diagnostic and Statistical Manual of Mental Disorders 41, 54–5, 134n
diagnostic categories 54, 55, 148
dialectical approach 47–8
disavowal 47
diversity (in psychoanalysis) 113–15, 121–2; danger in 118–20; impediments to 117–18; and pluralism 118–19 (*see also* pluralism); *see also* psychoanalytic schools
dreams 113

Eco, Umberto 140
ego needs 30; *see also* need gratification
emotion 123; twofold approach to 47; *see also* enthusiasm
emotional contagion 82
empathic expansion, impediments to 87
empathic failures 25
empathic understanding 95; *see also* understanding
empathy 25, 45–6, 77–8, 88–9, 97–9, 141–2; as antidote to rage 140–1; clinical considerations 84; components of 89; conceptualizations of 90; the continuum vs. the "magic moment" 79–80; and cruelty 45, 141, 142; definitions 45, 77–9, 89, 90, 106; developmental line of 78, 81–3; development of 80; and indifference 87–8; Kohut on 8, 10, 35, 45, 79, 89–90; mature 84; neurobiology of 44–5; oxytocin, discourse, and 80–1; psychoanalytic view of 86–7; types of 89–90 (*see also* self-empathy; sustained empathy); *see also* understanding
empathy circuit 44
empirical vs. hermeneutic studies 17, 19, 20, 22–4, 26, 27, 34, 37
enactivism 14, 71–2
enthusiasm 130–1, 133–4; and change 129–30; clinical examples 124–7, 132–3; depression and 124–5, 130–2; development of 123–4; and excitement 128; and meaning 128; and morality 129; and the person 128–9; psychopathology of 131–2; a tale of two artists 127; a variant of 133

eurociality, theory of 49–50
explanation(s): categories of 18; coexistence of multiple 21; distinguishing psychoanalytic pathology and 63–4; psychoanalysis and 26, 30; *see also* understanding
extrospection 19; vs. introspection 34–5, 101

Fearn, Nicholas 9–10
Ferenczi, Sándor 86, 88
field theory of mind 8
fore-conception 91
fore-having 91
fore-sight 91
fragmentation 137–9
Freud, Anna 53, 58, 86, 88
Freud, Sigmund 29, 39, 50, 99; on aggression 114, 141; diagnosis and 54; on dreams 113; on normality vs. pathology 53; psychiatry, psychoanalysis, and 3, 27, 144, 145

Gadamer, Hans-Georg 26, 31
Gallese, Victor 13
Gedo, John E. 13, 86
Gitelson, Maxwell 27, 33, 34
Glover, Edward 78
Goldberg, Arnold 8, 88
Goldner, Virginia 57
gratification 30, 33–4
Gross, C. 15
groups, rage in 140

Hagoort, P. 90
Hedges, D. 148
Heidegger, Martin v, 8, 10, 13, 91
hermeneutic circle 91–4
hermeneutic intervention, conditions that call for 34
hermeneutics 18, 19, 26, 113–14; defined 91; vs. positivism 30; psychoanalysis and 17, 19–21, 23, 26, 27, 30–1, 33; vicarious introspection as 35
hermeneutic understanding 19, 26
hermeneutic vs. empirical studies 17, 19, 20, 22–4, 26, 27, 34, 37
Holzman, Philip 19, 26, 40, 109
Home Alone (film) 70
homosexuality 58, 91–2
host vs. agent 55–6
Hume, David 9, 123

Index

Ingram, D. 31
insight 151
integration, psychological: criteria for assessing optimal 53
interpretation 37
introspection: vs. extrospection 34–5, 101; vs. observation 19; vicarious 35

Jacobs, David 55, 56
Jones, Ernest 53

Kandel, Eric 143
Kernberg, Otto F. 56
Klein, Melanie 53
"knowing" in children, development of 92
Kohut, Heinz 26, 58, 137; on aggression 136; on developmental deficits and goals of analysis 25; empathy and 8, 10, 35, 45, 79, 89–90; Heidegger and 8; on intuition 84; on narcissism and narcissistic pathology 61–2; selfobjects and 8, 9, 35, 115; on selfobject transferences 30, 61–2
Kubie, Lawrence S. 29–35, 37, 65
Kuhn, Thomas S. 32, 37
Kurzweil, R. 2, 67

language 9
lay analysis 39, 40
Levelt, W. 90
Lewis, Michael 82
Lindquist, K. 47
Lipps, T. 83
listening 30
locationist approaches to emotional categories 47
Luzzi, J. 18

MacCulloch, D. 36
MacIntyre, A. 117
Maturana, H. R. 71–2
maturity, emotional 53, 86
meaning: categories of 18; enthusiasm and 128; vs. facts/truth 18; multiplicity of 21; search for 18; truth and 46
mentalization 97; empathy and 82, 84, 97; see also theory of mind
Merleau-Ponty, Maurice M. 67
Metzinger, Thomas 96, 97
Michels, Robert 58
Miller, G. 80–1

mind: conceptions of the 8, 84 (see also theory of mind); extending beyond the brain 8; self, brain, and 8; see also "being kept in mind"
mind reading 9, 82–3
mind vs. brain debate 2, 7
Mitchell, Stephen A. 30, 34

narcissism and narcissistic pathology 61–2
narcissistic injury 139, 140
narcissistic rage 136; see also rage
need gratification 30, 33–4
need-wish dilemma 29–30
neurobiology: behavioral 47; see also neuroscience
neuropsychoanalysis 146
neuroscience 2, 3, 8, 24; see also under psychiatry and psychoanalysis
Nöe, Arva 11, 13, 67, 75, 81
nonlinear dynamic systems 88
normality 53; criteria for 53; definitions 53
normativity 53
norms 53–4; and psychiatry 54–6; and psychoanalysis 56–7

objectivity 18; vs. subjectivity 34–5
observation vs. introspection 19
Orwell, George 140
oxytocin, empathy, and discourse 80–1

patients: feeling cared for 29–30; wishes vs. needs 29–30
perspective and empathy 89
perverse behavior 126–7
Piaget, Jean 8, 69, 70
pluralism 2–3, 18, 26; "theoretical" 118–19; see also diversity
Popper, Karl 34, 36
positivism 30
pragmatism 23, 34
Preston, S. D. 77, 79
projective identification 86
prosopagnosia (facial blindness) 46
psychiatry, history of 1–2
psychiatry and psychoanalysis 39, 49–51; an exercise of difference 48–9; and the brain 46–8; contrasted 33; diagnosis in 41–2; language and dependence 40–1; neuroscience and reconciliation 44–6; treatment in 42
psychic powers 9

psychoanalysis: as an understanding vs. explaining psychology 26, 30; carving out a place for 143–9; compared with biblical hermeneutics 33; decline 1–2, 146; goals 18–19, 21, 26–7, 33, 151; independence 146–9; opportunity for 148–9; recognition 146–8; scientific status of psychotherapy and 29–37; therapeutic factors in 29–30; *see also specific topics*
psychoanalytic explanations and psychoanalytic pathology, distinguishing 63–4
psychoanalytic schools 18, 19, 26, 33, 119; deviant 118–20; distinctions between 115–17; *see also* diversity
psychoanalytic training 39–40
psychodynamic formulations 64
psychodynamic psychotherapy 24–5, 42, 50, 51; goals 18–19, 27, 151; vs. psychoanalysis 17–19, 26–7, 33; *see also under* psychoanalysis
psychopathology and psychoanalysis 56
psychopathy as indicator of knowing without feeling 84–5
psychotherapy *see* psychodynamic psychotherapy

rage 135, 139–42; case examples 136–9, 141; empathy as antidote to 140–1; phenomenology 136; psychoanalytic theory of 136–8; psychology of 136
Reik, Theodore 37
relational psychoanalysis 30
representations 69–71
research tool, psychoanalysis as 148
Ricoeur, Paul 45, 46, 81, 144
Rizzolatti, Giacomo 79
"road rage" 135, 140

Sabshin, Melvin 55
Sacks, Oliver 79–80
Schafer, Roy 79
science 32; hermeneutic vs. empirical 17, 19, 20, 22–4, 26, 27, 34, 37
self 2
self-definition 140
self-empathy 87, 101–2, 106–7; as antidote to helplessness 106; case reports 104–6; limitations 103–4; need for 104; time line 102; unconscious factors 102–3
self-fragmentation 137–9

selfobject failures 139; *see also* narcissistic rage
selfobject needs 30
selfobject relations 35
selfobjects 8, 9; Kohut and 8, 9, 35, 115
self psychology 8; *see also specific topics*
sexual orientation 58
sexual perversion 126–7
Sheldrake, Rupert 8, 9, 13
silence (and the silent analyst) 37; pragmatics of 35–6
silo effect 3
social sciences 32
Sohms, M. 47–8
Stepansky, Paul E. 118–20
Strawson, Galen 13
sustained empathy 89–97; case reports 91–4; meanings 90–1; prerequisites for 96–7; therapeutic effect 94–6
symptom relief 18–19, 27, 33

telepathy 9
theoretical orientations *see* psychoanalytic schools
theory of mind (ToM) 8, 82, 84, 89; *see also* mentalization; mind
Thompson, Judith Jarvis 56
Tomkins, Silvan S. 47
transference(s) 29, 36, 64–5, 86; abnormal 61–2
truth and meaning 46
Tuckett, D. 33
Turnbull, O. 47–8
Two Minute Hate 140

unconscious, neurobiology and the 143
understanding 17–22, 25–7; being understood and feeling understood 22–5; case material 20–4; empathic 95 (*see also* empathy); empiricism, neuroscience, and 24; explaining 26–7; Heidegger on 8; Kohut on 8; meanings and 26; psychotherapy and 24–5; types of 19

Varela, F. J. 71–2
vicarious introspection: as hermeneutics 35; *see also* introspection

Whitaker, Robert 55
Wilson, Edward D. 49–50
Wilson, M. 89
wishes vs. needs 29–30
Wittgenstein, Ludwig 89

Lightning Source UK Ltd.
Milton Keynes UK
UKOW06f1822060815

256523UK00003B/156/P